International Scientific and Technical Conference: Structuring, Strength and Destruction of Composite Materials and Structures (SSDCMS 2021)

Held online due to COVID-19

IOP Conference Series: Materials Science and Engineering
Volume 1162

Odessa, Ukraine
8 - 9 April 2021

Editors:

V.M. Vyrovoy
S.O. Kroviakov
S.S. Makarova
N.V. Kazmirchuk

ISBN: 978-1-7138-3429-8
ISSN: 1757-8981

Printed from e-media with permission by:

Curran Associates, Inc.
57 Morehouse Lane
Red Hook, NY 12571

CURRAN ASSOCIATES INC.
proceedings
.com

Some format issues inherent in the e-media version may also appear in this print version.

Institute of Physics
Dirac House, Temple Back
Bristol BS1 6BE UK

Phone: 44 1 17 929 7481
Fax: 44 1 17 920 0979

techtracking@iop.org

Additional copies of this publication are available from:

Curran Associates, Inc.
57 Morehouse Lane
Red Hook, NY 12571 USA
Phone: 845-758-0400
Fax: 845-758-2633
Email: curran@proceedings.com
Web: www.proceedings.com

TABLE OF CONTENTS

Abstract

We are pleased to present the materials of the International Scientific and Technical Conference "Structuring, strength and destruction of composite materials and structures". The conference was organized by the Odessa State Academy of Civil Engineering and Architecture together with Ariel University (Israel), University North (Croatia), Gheorghe Asachi Technical University of Iasi (Romania), State Scientific-Research Institute of Building Structures, LLC CLH «KAMBIO», LLC «Stikon», SPC «Ecostroy», SE «Budova».

The conference brought together scientists and practitioners to the platform to discuss the mechanics and physics of destruction of concrete and reinforced concrete, structure formation, modeling and optimization of composite building materials and structures, and analysis of the work of these materials in structures. Presentations were made by theorists dealing with the issues under consideration and practicing developers-builders who are extremely interested in introducing the results of scientific research directly into the construction industry. Due to the Covid-19 pandemic, the event was held online on April 08, 2021. The conference was attended by about 60 participants.

We hope that future conferences will be just as successful and inspiring, as the issues raised at them are of great practical importance. The organizing committee of the conference expresses gratitude to all participants for their cooperation and indifference.

Head of the conference organizing committee, Rector of the Odessa State Academy of Civil Engineering and Architecture, Anatoly Kovrov

Conference materials are available in pdf format.

List of committees:

Kovrov Anatoly, PhD, Professor, Rector of OSACEA, Vice-President of Energy Academy of Ukraine, Head of the territorial department of Construction Academy of Ukraine, Chairman, Head of Organizing Committee
Vyrovoy Valeriy, Doctor of Engineering Sciences, Professor, Honored Worker of Science and Technology of Ukraine, OSACEA, Deputy Chairman of the organizing committee;
Kroviakov Sergey, Doctor of Engineering Sciences, Assistant Professor, Vice-rector, OSACEA, Deputy Chairman of the organizing committee;
Zavoloka Michael, PhD, Assistant Professor, Head of the Department of Building Structures Production, OSACEA, Deputy Chairman of the organizing committee.

Editors

Editor-in-chief – Vyrovoy V.M. – D.Sc., Professor, Odessa State Academy of Civil Engineering and Architecture; 4, Didrikhsona str., Odessa, Ukraine, 65029. Tel. +38 097 498 49 79 vyrovoy@ukr.net
Scopus ID: 57192371695, https://orcid.org/0000-0001-8818-4112;

Executive editor – Kroviakov S.O. – D.Sc., Associate Professor, Odessa State Academy of Civil Engineering and Architecture; 4, Didrikhsona str., Odessa, Ukraine, 65029. Tel. +38 048 723 49 37,
+38 067 922 18 90. skrovyakov@ukr.net
Scopus ID: 57215915002, https://orcid.org/0000-0002-0800-0123;

Coordinator – Makarova S. S., Ph.D., Professor, Odessa State Academy of Civil Engineering and Architecture; Department of Production of Building Products and Constructions, Staroportofrankovskaya st., 28, Odessa, Ukraine, 65020.
Tel. + 38 096 559 39 98, svetlana.unay@ukr.net

Coordinator – Kazmirchuk N. V., Ph.D., Senior Lecturer, Odessa State Academy of Civil Engineering and Architecture; Department of Production of Building Products and Constructions, Staroportofrankovskaya st., 28, Odessa, Ukraine, 65020.
Tel. + 38 067 07 49 98, bti.vbk@ukr.net

IOP Conf. Series: Materials Science and Engineering 1162 (2021) 011002 doi:10.1088/1757-899X/1162/1/011002

Peer review declaration

All papers published in this volume of *IOP Conference Series: Materials Science and Engineering* have been peer reviewed through processes administered by the Editors. Reviews were conducted by expert referees to the professional and scientific standards expected of a proceedings journal published by IOP Publishing.

- **Type of peer review:**

  For each article, two reviews are presented (the first is internal, the second is external). The peer review is open.

- **Conference submission management system:** e-mail

- **Number of submissions received:** 25 manuscripts

- **Number of submissions sent for review:** 25 manuscripts

- **Number of submissions accepted:** 21 manuscripts

- **Acceptance Rate (Number of Submissions Accepted / Number of Submissions Received X 100):** 84%

- **Average number of reviews per paper:** 2 per paper

- **Total number of reviewers involved:** 36 reviewers

- **Any additional info on review process:**

  Manuscripts were sent to the conference organizing committee. Then the members of the organizing committee sent these manuscripts to the reviewers. Each manuscript was reviewed by two reviewers - candidates of technical sciences and doctors of technical sciences. The following requirements were imposed on the articles: clarity of presentation, relevance of the topic, originality and novelty of the idea, correspondence of the material to the declared topic of the conference, transparency of the result obtained. If the reviewers accepted the articles or pointed out deficiencies requiring changes, the organizing committee sent the manuscript back to the authors so that the authors could make the necessary corrections and changes. After making corrections to the article, the authors sent the final version of the manuscripts to the organizing committee. After re-examination, the editors completed proofreading of the manuscripts to fully comply with IOP Publishing requirements.

- **Contact person for queries:** Svetlana Sukhanova, svetlanasukhanova@rambler.ru

IOP Conf. Series: Materials Science and Engineering 1162 (2021) 012001 doi:10.1088/1757-899X/1162/1/012001

Nanomodification of mineral binders

V Derevianko[1a], **N Kondratieva**[2b], **V Volkova**[3c], **H Hryshko**[4d]

[1] Department of Technology of Construction Materials, Products and Designs, Prydniprovs'ka State Academy of Civil Engineering and Architecture, 24-a Chernyshevsky St., 49600, Dnipro, Ukraine, ORCID ID: 0000-0003-4131-0155
[2] Department of Chemical Engineering and Binding Materials, Ukrainian State University of Chemistry and Chemical Technology, 8 Gagarin prosp., 49005, Dnipro, Ukraine
[3] Department of Operation of Hydromelioration Systems and Construction Technology, Dnipro State Agrarian-Economic University, 25 Serhii Efremov St., 49600, Dnipro, Ukraine, ORCID ID: 0000-0002-1883-1385
[4] Department of Operation of Hydromelioration Systems and Construction Technology, Dnipro State Agrarian-Economic University, 25 Serhii Efremov St., 49600, Dnipro, Ukraine, ORCID ID: 0000-0001-7046-1177

E-mail: [a] viktorderevianko2017@gmail.com, [b]nataliyavk@yahoo.com, [c]drvev09@gmail.com, [d]gryshko.anna0101@gmail.com

Abstract. In the result of researches performed on the directed regulation of properties of gypsum binders, it was determined that the specific surface of various anisotropic surfaces regulates the shape of crystals as well as kinetics of their formation and growth. It has been established that the process of hydration and formation of gypsum binders can be regulated through the size of the specific surface, parameters of nanoadditives as well as through the change of concentration in solution. The influence of micro- and nanoadditives on the change of the system surface energy, as well as on introduction of plasticizer and change of time solution is in the supersaturated state (which increases the number of nuclear centers of crystallization) has been established experimentally and theoretically proved.

Nanotechnology is the science of direct influence on individual atoms and molecules and the main purpose of research is to create structures of materials with defined properties. The essence is in the fact that particles with nanomaterial dimensions have special properties, which sometimes are not subordinated to the existing laws of physics, chemistry. Analysis of scientific achievements shows that the world is on the threshold of the third scientific and technological revolution, which is going to determine the place of countries in the technological sphere of the XXI century. Up to date, more than 70 countries are implementing state programs for nanotechnology development. More than 29000 scientific centers and "nanocompanies" are organized and successfully functioning in the world. They carry out research and have certain developments in the field of nanotechnology and related disciplines.

There is no international information on financing nanotechnology in Ukraine. It is the more so in the construction industry. For the sake of justice, it should be noted that the research is being conducted. Example: Kharkiv Machine Building Plant (FED Plant), Malyshev Factory, "Motor Sich" in Zaporizhia, Kharkiv Institute of Railway Engineers' Medical Center.

So a research laboratory and NANONAT-Eps workshop are functioning at the FED Plant. At PSACEA (2015), KNAHU (2018), KNUCA (2020) doctoral theses were defended. Their authors considered theoretical issues concerning application of nanostructures for modification of composite materials. Ukraine has signed a project to support the development of nanotechnologies in Europe (European Partnership countries (EPS), Institute for Problems of Material Sciences (IPMS). The pilot project Europe: Armenia-Germany, Belarus-Luxembourg and Ukraine-UK. NANONAT-EPS - November 2013 Kyiv (Ukraine).

The relevance of the research is characterized through the data of financial costs for the development and manufacture of products with the use of nanotechnology (Fig. 1): 2008 - USD 200 billion, 2015 - about USD 1,000 billion, 2020 – more than USD 2,000 billion [1].

The main trends of nanotechnological development in the field of construction are directed on designing new products, with improved quality and functional characteristics, as well as on increasing application efficiency of existing materials [2]. Materials have unique properties and their use can make a revolution in many spheres of construction materials production. [3].

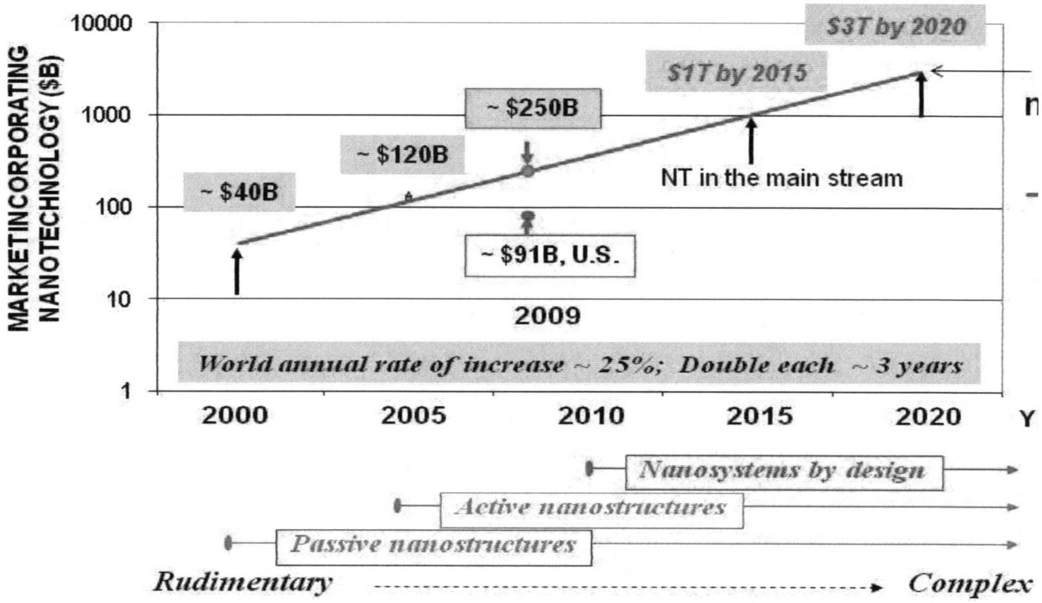

Fig. 1. Scheme of marketing for production with the use of nanotechnologies

With the help of nanoparticles you can change the color of coating, create self-cleaning functions, form a special antibacterial layer and perform ultraviolet protection. Structural composite materials based on nanotechnologies have a strength that many times exceeds that of traditional analogues.

Developments in the field of construction nanotechnologies of building products are distributed in the following way. The largest segment belongs to nanomodifications of cement binders (Fig. 2).

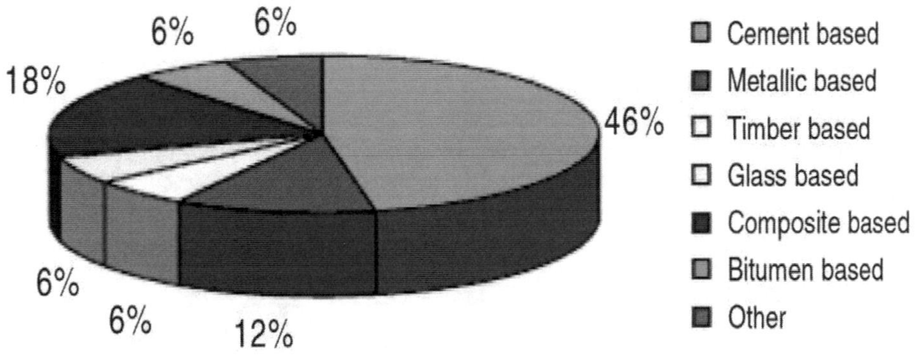

Fig. 2. Distribution diagram concerning modifications of construction products

In this concern a promising area is the use of nanomaterials for the modification of gypsum binders. The most widespread hypotheses explaining the change in properties of binders during modifying by means of ultradisperse additives and nanoagents can be grouped as follows: Nanoparticles cause influence on the electrosurface properties (Fisher G B and Plugin A A); rheological characteristics are changed (Voitovich Y V); nanostructures play the role of crystallization centers (Yakovlev and Derevianko V M); the number of point contacts is increased (Gharkavi M S and Yakovlev); the effect of introduced nanoparticles is associated with an additional active specific surface area of the phase separation and excessive surface energy Buryanov A F, Tokarev Y V, Sanitskiy M A and Tolmachov S M).

It should be noted that the concept of phase is expressed less clearly in conditions of reduced particle size. It is difficult to determine borders between homogeneous phase and heterogeneous phase and between amorphous and crystalline states. The usual physical and chemical ideas, including notions of "composition-property", "structure-function" are in this case supplemented with such terms as "size", "self-organization" [3-4].

Introduction of binding nanoparticles into the mixture results in a change of structure formation and affects the properties of construction materials. The influence of dimensional effects (the upper limit is considered to be particle sizes up to 100 nm) causes a change in physical and mechanical properties in the specified size range of materials [3]. The Nanoparticles are classified: by size, energy and geometric characteristics, composition, shape, methods of formation [5]. The following types of nanostructures are most often distinguished by their shape: fullerenes, single-layer and multi-layer nanotubes, nanowires, hemispheres and nanoparticles, nanoclusters and nanocrystals.

In our studies, we used carbon nanoparticles (CNP), carbon nanotubes (CNT), functionalized carbon nanotubes (CNT-OH), and carboxylated carbon nanotubes CNT-COOH) Gypsum plaster was used as a binder Plasticizers: MasterAir 81 (fig.), Master X-Seed 100, Master X-Seed 100, Glenium ACE 430, Polyplast P1, Sika Retardatcur.

Application of carbon nanotubes (CNT) for modifying properties of gypsum binders is considered as an effective means of improving the physical and mechanical properties of composite materials.

Improved efficiency of carbon nanomodifier in composites can be achieved by means of chemical modification of the CNT surface by functional groups, for example hydroxyl group or carboxylic group [6] (Fig. 3).

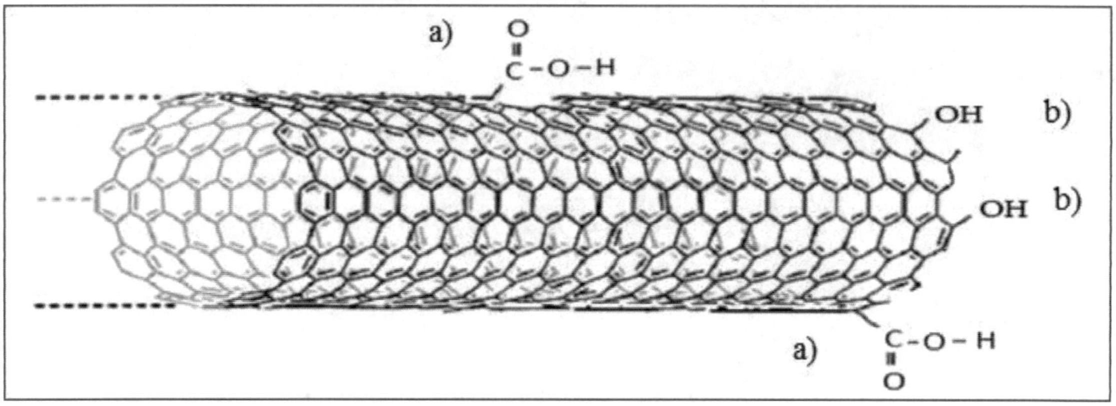

Fig. 3. The scheme of functionalization by means of CNT groups is as follows:
(a) carboxyl Groups; (b) hydroxyl Groups

Carboxylation of carbon nanotubes was carried out by interaction of the latter with various oxidizing agents [4, 7] (nitric acid in various media, salts of chromium and manganese in higher oxidation states, hydrogen peroxide). For CNT oxidation with hydroxyl groups a mechanochemical method is used, which consists in a joint grinding of CNTs and alkali during 60 minutes. Figure 4 shows CNTs after ultrasound processing. Functionalized carbon nanotubes (unlike initial ones) do not form large agglomerates and are well separated.

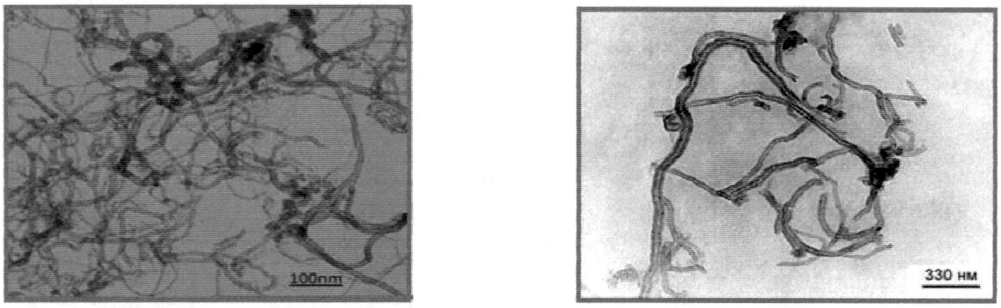

a) b)

Fig. 4. Photomicrographs of multilayer carbon nanotubes: (a) – pure CNTs; (b) – modified CNTs.

Modification of binders with nanosystems is complicated due to obtaining a heterogeneous structure (because of small number of nanocomponents). Technology of their introduction into the composition of dry binders is in the stage of development, and at present the process includes preparation of a dispersed solution "nanosystem-sealant". In order to reduce sedimentation, increase the viability of the solution and form heterogeneity of the structure studies were conducted with various sealants [7] (Fig. 5). Dispersion was performed with the use of ultrasound in the following media: CNT - water; CNT – plasticizer - water; CNT-OH – plasticizer - water; CNT-COOH - plasticizer - water.

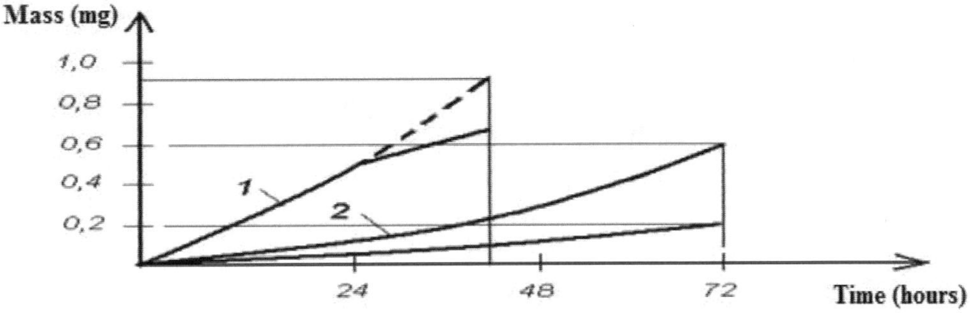

Fig. 5. Graphs of integrated curves reflecting sedimentation of nanomodifier particles: 1) CNT - water; 2) CNT- plasticizer - water; 3) CNT-OH - plasticizer - water

For studying the influence of nanoadditives on the processes of crystallization the method of counter diffusion of crystals growing [8, 9] has been improved. According to the stereochemical reaction:

$$CaCl_2 + Na_2SO_4 + 2H_2O \rightarrow CaSO_4 2H_2O + 2NaCl,$$

prepared solutions Na_2SO_4 and $CaCl_2$ were added in different containers with distilled water (Fig. 6). Concentrations of solutions were calculated taking into account the maximum solubility of these substances at room temperature. Crystals are formed through diffusion in an aqueous medium at room temperature. In general, experimental studies of the effect caused by nanosystems on the structure and properties of gypsum binders were carried out according to the following scheme, determining the shape and other parameters of crystals grown based on $CaCl_2, Na_2SO_4, 2H_2O$: 1 solutions - without additives; 2 - with addition of plasticizers; 3 - with addition of plasticizers and ultra- and nanosystems (nanoparticles, nanotubes). In parallel, samples based on gypsum plaster of similar compositions were used to test the basic physical and mechanical properties.

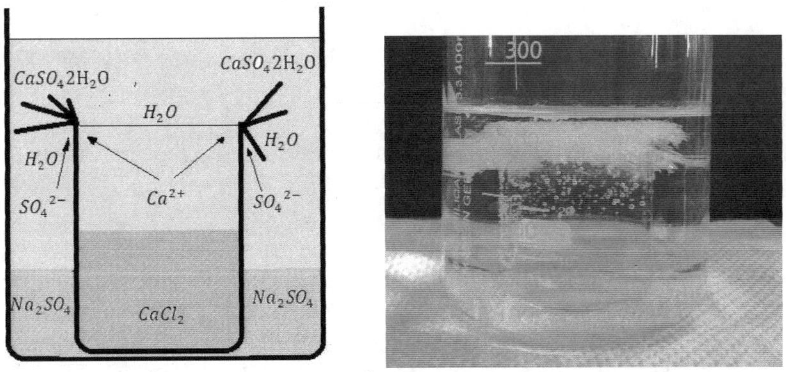

Fig. 6. Scheme of growing crystal on the basis of counter diffusion method with the use of *$CaCl_2$ and Na_2SO_4*

The choice of initial materials was based on the proposed scientific hypothesis consisting in the fact that mineral binders with special properties can be obtained by means of controlling processes of hydration and structural formation of crystal lattice with necessary characteristics while introducing nanomodifying additives.

The block of initial materials consisted of mineral binders (variety of gypsum plaster) and surface active substances (surfactants), nanoadditives (in the form of 1D-3D nanoparticles of CNT (fullerenes), CNT nanotubes, dispersive solutions of industrial waste and natural materials), plasticizers [10]. Calcium chloride and sulfates were used for the growth of calcium sulphate dehydrate crystals.

The purpose of the research is to study the effect of nanomodifiers on the mechanism of the hydration process and structure formation of gypsum binders.

Processes of hydration of binders are directly related to the subsequent processes of crystal formation. Their structure forms and sizes determine the properties of hardened materials. Results of growing gypsum crystals from saturated solutions of $CaCl_2$ and Na_2SO_4 (using the method of counter diffusion without/with ultra- and nanoadditives are presented below.

Theoretical and experimental substantiation of composite formation principles and peculiarities of hydration processes, structure formation of gypsum systems as well as analysis of their modification possibilities has been carried out.

Based on the technological chain (raw materials - thermal treatment - hydration processes - hardened structure) it is possible to emphasize that the process of hydration is the most manageable, although it is also more complicated from the theoretical point of view. All modifications of gypsum systems are based on calcium sulfate dehydrate ($CaSO_4$-$2H_2O$). It has a very perfect degradation in the plane (010), perfect degradation in the plane (111) and imperfect degradation in the plane (100). There are two different types of water: weakly retained (1.5 H_2O) molecules are attached directly to calcium and strongly retained ones ($0.5H_2O$) are bound to hydrogen sulfide ion) (Fig. 7). 7). Modification is aimed at changing the structure [11].

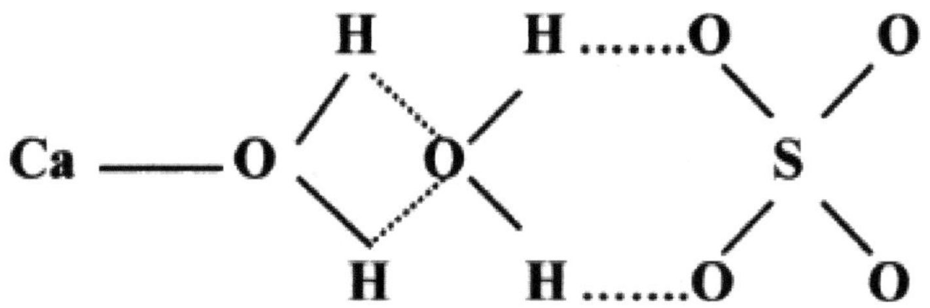

Fig. 7. The structure of relations in calcium sulphate dehydrate

Energy calculations of modification strips have the following values: Ik $SaSO_4$ $2H_2O$ q 2300 kJ/mol; Ik $SaSO_4$ $0{,}5H_2O$ q 2505 kJ/mol; Ik $SaSO_4$ 2638 kJ/mol. Analysis of various structures of gypsum crystals shows that the key factor is the change of the system energy state [11].

The strength of the structure formed is largely determined by the size and saturation kinetics in the suspension liquid phase. Therefore, change in hydration process conditions can regulate formation of structure, framework and properties.

Given the model, the faces on the crystal surface have different structures at the atomic level, depending on their orientation. There are three types as functions of the number of parallel lines of connections; F (2 or more connections), S (1 connection) and K (no connections) [12, 13].

Control of the structure formation process is determined primarily by the degree of mineral hydration, the number and type of crystal intervals between them depending on the degree of liquid phase saturation relative to the maximum dissolution of hydrates crystallized from it.

Adsorption of modifiers is possible: – on the edges of crystals, which have the largest surface energy. For example, polymers adsorb mainly on the edges of {111} and {011}, and thus they change the rate of crystal nucleation [14].

The article presents the results of studies of the effect of ultra-plasticizers and nanoadditives on hydration and structural processes. They represent a complex that combines chemical interaction and formation of hydrate modifications, and accompanies physical, physical and chemical and colloidal processes. The technology of modification of binders with nanoadditives, according to the results of previous tests, is not yet possible without the use of PAW. In this regard, the study was initially conducted concerning their impact on hydration processes and the structure of gypsum systems.

IOP Conf. Series: Materials Science and Engineering 1162 (2021) 012001 doi:10.1088/1757-899X/1162/1/012001

Adding a PAW to the calcium sulphate hemihydrate solution changes the rate of physical and chemical processes: electrical conductivity, thermodynamics, rate of hydration processes (Table 1), and morphology of crystals (Table 2) and structure of solutions.

In the case of growing crystals in a medium without a plasticizer, we can say that the decisive factor is the degree of solution re-saturation and the time of the system. Introduction of a plasticizer into the crystal-forming solution (in addition to changed degree of re-saturation and time in the pre-saturated state) passes through the stage of plasticizer selective adsorption on the crystal faces. The surface energy is changed in the result.

Table 1. Change in the duration of calcium sulfate hemihydrate Hydration Processes

Properties	Without additives	Master 81	M-X-seed 100	Gleneum ACE 430	Sika Private
V/G ratio	0.62	0.75	0.67	0.52	0.49
$T_{si.}$– time of setting initiation, min/sec	10 / 10	12 / 30	12/ 30	13 / 30	18
T_{fs} – time of final setting min/sec	14 / 50	13 /35	13 / 05	20 / 10	35

Table 2. Morphology of Crystals

Surfactants	Microphotography of crystals	Schematic figures	Crystal dimensions, μm	
without a plasticizer		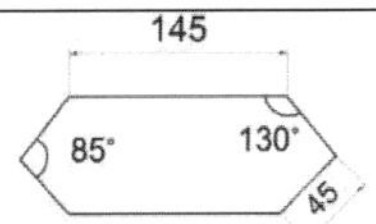	145	0
MasterAir 81		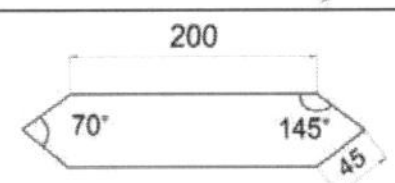	200	5
Master X-Seed 100	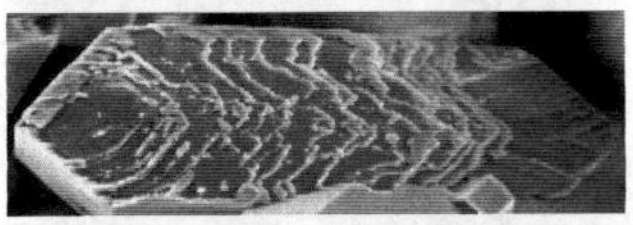	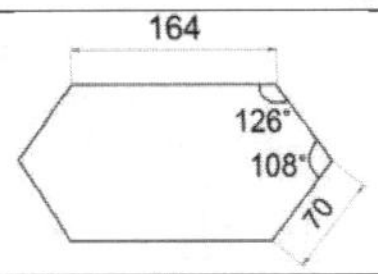	164	0
Glenium ACE 430	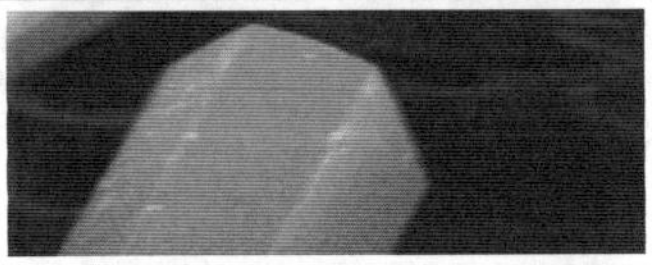	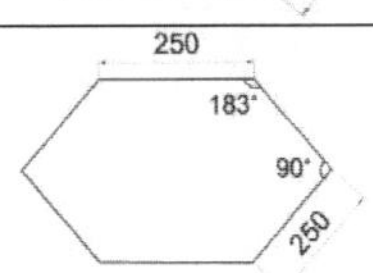	250	50
Sika (Ivate)		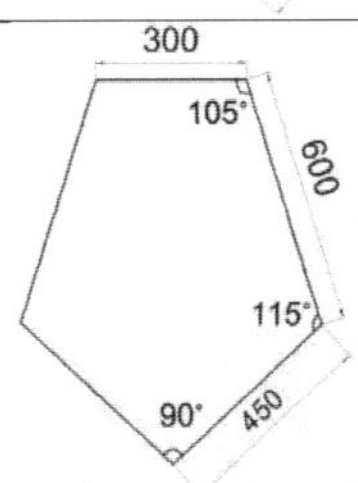	300	50

Adsorption on the faces of crystals plays an important role in the process of their growth. Selective absorption leads to the blockade of growth along the axis (molecules are adsorbed on the upper part of the face) and there is a tendency of growth with a low aspect ratio, and as a result crystals are formed as thickened columns.

In order to study the influence of the crystal-forming medium on the morphology of calcium sulphate dehydrate crystals they were grown using the method of counter diffusion without additives and with the addition of plasticizers. According to the results it has been shown that under identical growth conditions crystals change their shape and size depending on the type of plasticizer added. Crystals grown in the medium without adding a plasticizer are identical in their shape and size; they are evenly distributed on the lining with a thickness of 1 mm. Length of crystals is 3 mm, their width is 0.2 mm and their thickness is 0.06 mm.

Fig. 8. **Photomicrographs** of gypsum crystals, grown in the following ways: a) – without a plasticizer; b) – **and** with MasterAir 81 additive

Crystals grown in a medium with PAW: MasterAir 81 (Fig. 8), Master X-Seed 100, Master X-Seed 100, Glenium ACE 430, Sika (Fig.) change their shape, size, packing density (Fig. 8, 9).

Crystals grown in the presence of Sika plasticizer have a thickness from 0.7 to 1.5 cm, and the crystal width is 0,8-1 mm. It can also be observed that the growth of crystals begins from one point; possibly it is due to a change in surface tension. In general, all crystals formed arcuate sprouts. These crystals are larger than those grown with other plasticizers.

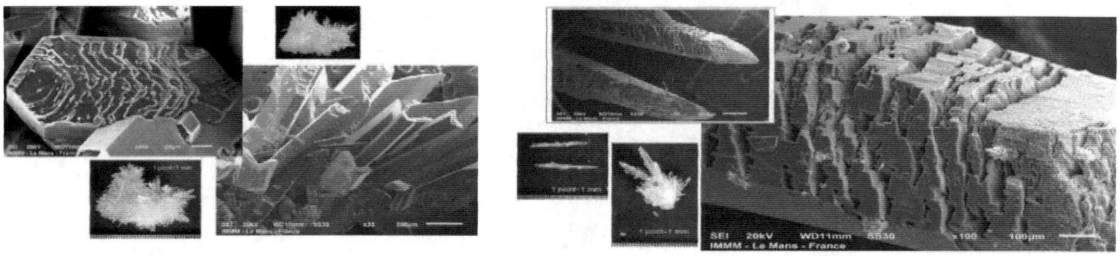

a) b)

Fig. 9. Gypsum crystals grown with addition of: b) – Master X-Seed 100; b) Sika

In photomicrographs you can see the growth layers, you can also see how the voids between the layers grow and the crystal is compacted, i.e. the crystal growth begins at various points on the surface, these layers begin to grow before meeting other layers.

The essence of modification processes in case of using ultra- and nanoadditives: It consists in the changed energy state of the system, which can be characterized through the ratio of components, the specific surface area of interface, temperature, pressure, solubility, diffusion etc.

Hydration reaction of samples with additives (Fig. 10, 11) is more intense; the time of reaching **the temperature maximum is less than that without additives.**

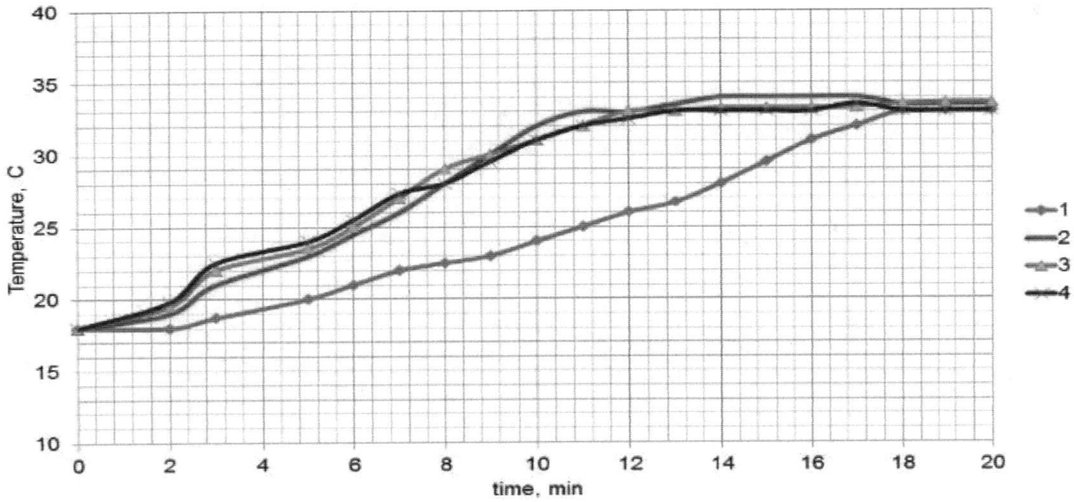

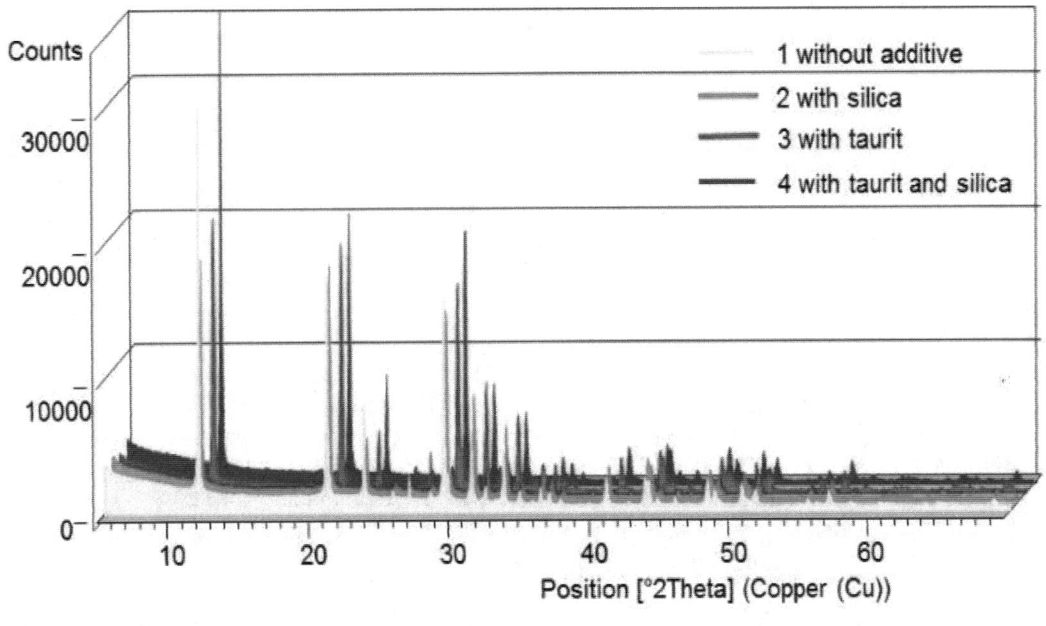

Fig. 10. Graphs of changes in temperature of gypsum hydration process a): with the addition of Taurit (0.5%) (2), carbon white (0.5%) (3) and with a complex composition of Taurit (0.5%) and carbon white (0.5%) (4). X-ray Images of gypsum: b): without additives (1), with silicon dioxide (2), with Taurit (3), with addition of Taurit and silicon dioxide (4)

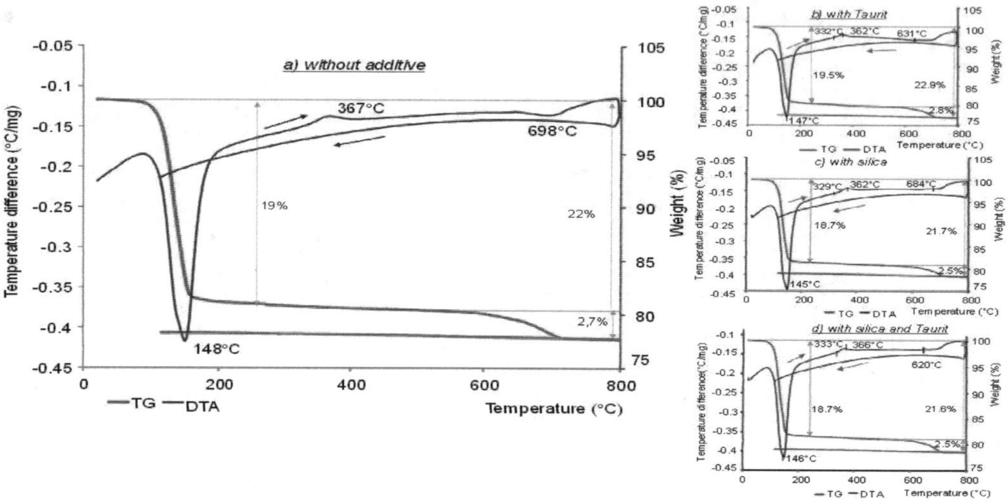

Fig.11. Results of the binding plaster DTA-TG without additives (a) with Taurit (b), with silicate (c) and with addition of Taurit and silicate complex (d)

The X-ray study of samples with additives shows that there are differences in the products of hydration processes. So the Taurit additive (Fig. 11b) reduces intensity of d q 0.756, 0.379 nm peaks, while intensity of d 0.286 nm peaks ($CaSO_42H_2O$) is increased. Taurit samples contain a lesser amount of $CaSO_4{\cdot}2H_2O$, then samples without additives, 2.5% $CaSO_4{\cdot}0.5H_2O$ i 1.0% $CaSO_4$ and they will not participate in hydration reaction.

Similar results are observed in the diffraction of X-ray powder (Figure 5c) with silicon dioxide plaster (silica plaster). Intensity peaks d q 0.757 ($CaSO_4{\cdot}2H_2O$) i d 0.378; 0.305 nm ($CaSO_4$) do not work and peaks d q 0.286; 0,267 ($CaSO_4{\cdot}2H_2O$). Peak d 0,348 ($CaSO_4{\cdot}0,5H_2O$) has almost gone. Slightly more than (1.7%) $CaSO_4{\cdot}0.5H_2O$ and (1.1%) $CaSO_4$ was not converted to dehydrate.

Studies of the changes in electrical conductivity and heat release intensity of Calcium Sulfate Hemihydrate in the process of its hydration have shown that the curve has three typical areas (Fig. 12). The first one corresponds to saturation of the solution by ions of calcium and sulfate. The second one is related with calcium dehydrate crystallization onset. The third area indicates the completion of the calcium sulphate dehydrate structural formation.

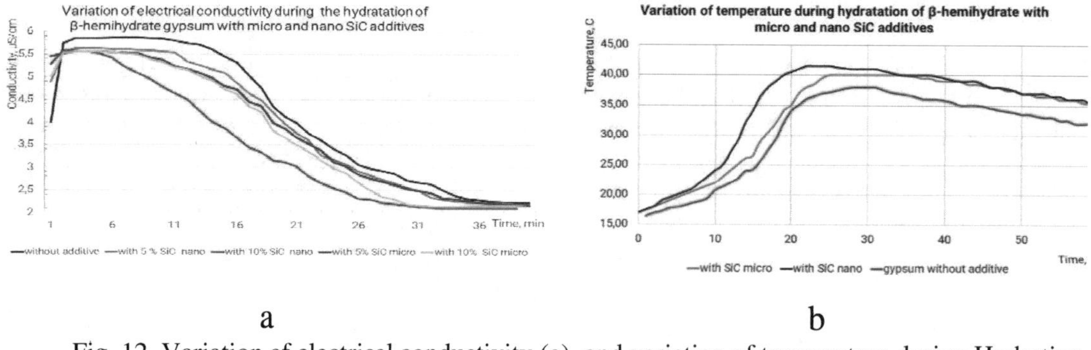

a b

Fig. 12. Variation of electrical conductivity (a), and variation of temperature during Hydration of β-Calcium Sulfate Hemihydrate with micro- and nano SiC additives (b)

According to studies of equilibrium morphology, surface stability is defined as a function of the directional distribution of the energy connection [14]. Adsorption on the surface of gypsum crystals depends on the charge density as well as on the distance between adjacent groups of ions. Gypsum crystals have a layered structure with SO_4^{2-} densely distributed on the side face and Ca2 + cations densely distributed in the upper part. Crystal faces {010}, {120}, {011} and {111} are arranged in the order of morphological value decrease. Change in the degree of solution saturation level, plasticizer adsorption on the faces as well as changes in surface energy affect the morphology of crystals.

At increase of microadditives presented in the solution up to 10% thickness of crystals reaches the level of 3 - 5 mm (Fig. 13), and their length grows up to 10-15 mm. The growth layers are more ordered, they are denser than those in crystals with lesser amounts of microadditives.

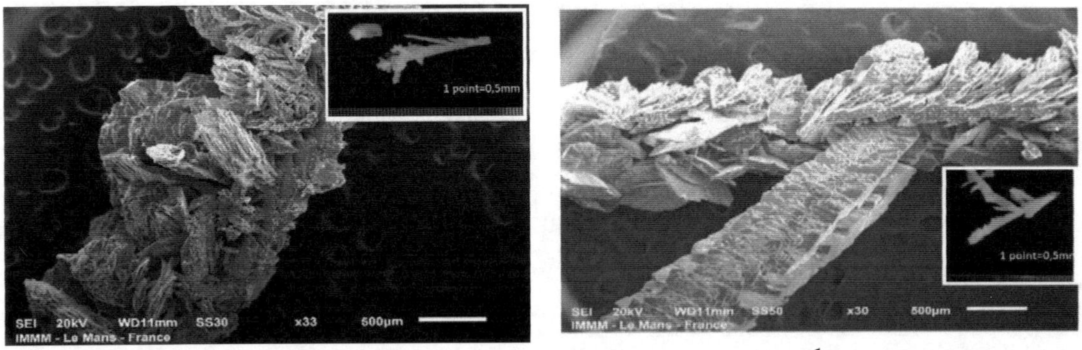

a b

Fig. 13. Photomicrographs of gypsum crystals grown in the following ways: a) – with a SiC microadditive (5% solution) and Sika superplasticizer (0.2%); b) – with a SiC microadditive (10% solution) and Sika superplasticizer (0.2%)

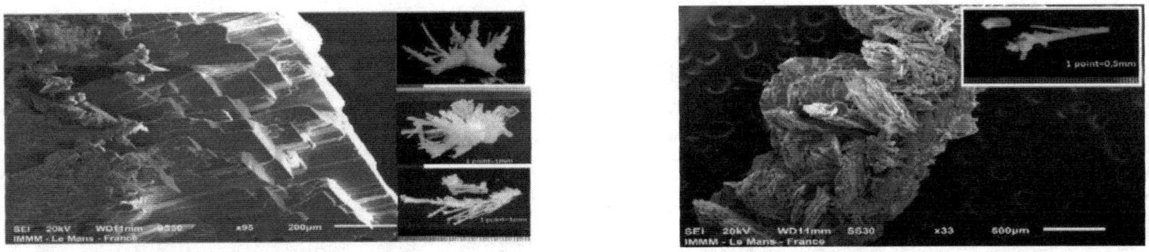

a b

Fig. 14. Photomicrographs of gypsum crystals grown with the following additives to the crystal forming medium: a) – SiC nanoadditive (5%) and Sika superplasticizer; b) – SiC nanoadditive (10%) and Sika superplasticizer (0.2%)

Theoretical and experimental studies of the influence caused by the specific surface of the components by the example of Calcium Sulfate Hemihydrate (G-5-N-II) and finely ground gypsum (G-5-N-III) gypsum indicate a change in the system energy state which is the driving factor of structure formation. So, granulometric composition of gypsum plaster sized up to 10 **μm is about 10%, and in** the finely ground one it is **25%.** At that the ratio of surface energy (M.A. Sanitskyi as the ratio of the surface area of particles to their volume, S / V) is increased by 3-5 times.

The study of optical density and spectral directional transmission coefficient ("Expert-003" photometer) of saturated solutions indicate a sharp change in optical properties, which is associated with the appearance of the solid phase.

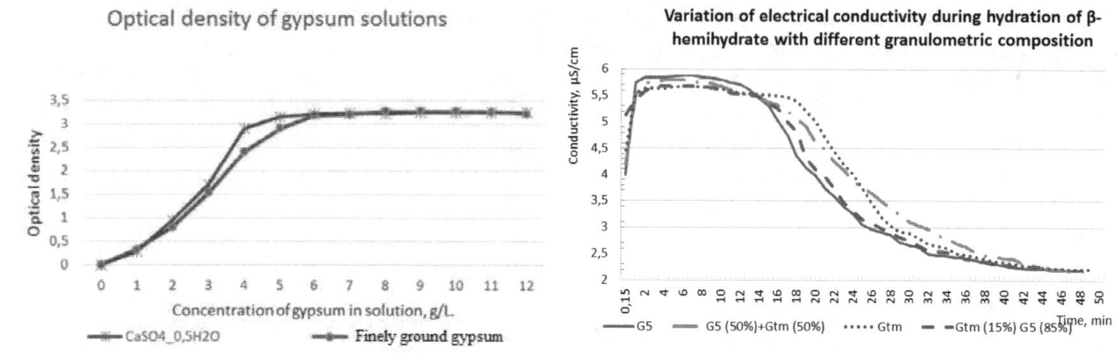

a b

Fig. 15. Graphs of changes (variations): a) – Variation of optical density of gypsum solutions with different sizes of factions; b) – Variation of Electrical Conductivity

The spectral directional transmission coefficient (nk) in an aqueous solution of gypsum and finely ground gypsum with a change of up to 5% in the range of 99.69% to 2.63 and 1.01, respectively. Increasing the concentration of components to 10% causes reduction down to 0.06, and down to 0.7 in finely ground gypsum, i.e. more than 10 times. Change of electrical conductivity takes place in the same way (Fig.15).

Theoretical and practical studies show the influence of faction specific surface (energy of systems) on the processes of hydration and structure, which gives an opportunity to change properties of the product.

Conclusions

According to the proposed hypothesis 'Regulation of crystal morphology and structure of mineral binders by means of additives with nanoparametric characteristics' as well as according to theoretical and experimental studies performed the nature of the influence caused by individual components of the hydration process has been determined.

1. Theoretical and experimental studies of structure formation processes during hydration of mineral binders have established the possibility of modifying their structure to achieve maximum strength. This strength depends on the degree of hydration of minerals, the number and type of crystal gaps between them, depending on the degree liquid phase saturation relative to the maximum solubility of hydrates.

2. An analysis of the effect caused by additives on the growth rate of certain crystal faces (which leads to a change in the properties of the solid body) has been performed. It has been found that there is a relationship between the directional growth of crystals and change in matrix characteristics depending on the concentration of Ca^{2+} ions in solution.

– Changes in temperature, pressure, pH of the medium, water-gypsum ratio in the process of hydration and hardening cause changes in structural topology and rate of crystal formation.

3. Mechanisms of modification of mineral binders have been compared by means of using ultrafine additives and nanoadditives. It has been established that the difference between the results as well as the lack of clear patterns can be explained through the following factors: Nature of modifiers, parameters of synthesis, method of introduction, surface chemical composition.

4. Dependence between the growth of crystals and variable of matrix characteristics (i.e. the directed regulation concerning properties of gypsum binders) has been established;

5. The following facts have been determined: the specific surface of various anisotropic surfaces is a key parameter of growth, which regulates the shape of crystals as well as kinetics of their nucleation and growth. Introduction of nanoparticles dramatically increases the value of surface energy which in its turn influences the process of hydration by means of increasing the number of crystallization centers

and the rate of the process as well as by means of changing the direction of growth (as indicated by hydration process thermograms B-$CaSO_4$-0.5H) and solution electrical conductivity;

It has been established that the process of hydration and formation of gypsum binders can be regulated through the size of specific surface area, parameters of nanoadditives as well as through the change of concentration in solution.

6. By means of using the method of counter diffusion for growing crystals of saturated solutions of $CaCl_2$ and Na_2SO_4 without/with additives the following aspects have been determined: crystals grown without additives are thin and needle-shaped, 3-5 mm long, 0.1-0.15 mm wide and 0.05 mm thick. And, for example, in the result of introducing SiC nanoadditive to the crystal-forming medium, the length and thickness of crystals increases: their length grows up to 5-8 mm, and width grows up to 0.1-0.8 mm. Adding SiC to the crystal-forming medium leads to growth of crystals on the C axis (microadditives of 5-7mm in length and 7-9 mm in width);

7. Based on studies aimed at determining coefficients of optical density and light transmission the influence caused by change of the surface energy of gypsum granulometric structure and finely ground gypsum as well as that of their mixture on properties of solutions has been established. The spectrum of directed light transmission and optical density depends on gypsum concentration in the solution. The change of gypsum electrical conductivity will be with different granulometric composition;

8. It has been established that: the change of the stage of saturated solution changes plasticizer adsorption on the faces of crystals as well as the surface energy, which are the main factors of influence caused on formation of crystals and on the process of crystal structural formation in general.

References

[1] Dementiev V 2017 *Struggle for Nanotechnological Leadership USA, EU, China, Russia* Moscow: CEMIRAS pp 1097–118
[2] Belichenko Y, Tolmachev S, Misko T and Duka A 2011 Features of the influence of carbonaceous nanoparticles on the rheological properties of cement paste and technological properties of the fine-grained concrete *Concrete and Reinforced Concrete in Ukraine No 6* pp 2-8
[3] Lina Kondofesky-Mintova. Fundamental Interactions between Multi-Walled Carbon Nanotubes (MWCNT), Ca^{2+} and Polycarboxilate Superplasticizers in Cementitious System *Proceedings Tenth International Conference* Prague, Czech Republic, October 2012 p 423-434
[4] Rybalkina M 2005 *Nanotechnology for Everybody. Big in Small* Moscow: Nanotechnology News Network p 444
[5] Kodolov V I and Khokhriakov N V 2008 *Chemical physics of the processes concerning formation and transformation of nanostructures and nanosystems. Tutorial* Izhevsk: ISTU p 704
[6] Derevianko V 2018 Study of impact factors and the mechanism of process of gypsum binder hydration in the presence of nanomodifiers *French-Ukrainian Journal of Chemistry (Electronic Materials* **6 (1)** Kyiv: Taras Shevchenko National University of Kyiv Web of Science in chapter Emerging Sources Citation Index pp 92–100
[7] Derevianko V N 2018 Impact of granulometric composition on mineral binder hydration processes *Journal of Engineering Science (Electronic Materials* **XXV(3)** Moldova: Publishing House Tehnica UTM pp 74–84
[8] Sanitsky M, Fischer H, Soltysik R and Korolko S 2003 The influence of modifiers on the gypsum binders' hydration *Crystal growing. Int. Conf. 15 (Conf. Materials* **1** Weimar: Ibausil pp 211–219
[9] Kondratieva N, Sanitsky M and Soltysik R 2017 Microstructure and properties of modified gypsum systems Weimar Plaster Conf. *Weimar: Bundesrepublik Deutschland tagungsbericht* ISBN: 9783000554476, 3000554475 p 162–165
[10] Derevianko V 2020 Modeling the Mechanism of Mineral Binders Hydration Processes in a Macro-Micro-Nanosystem *Nanosystems, Nanomaterials, Nanotechnologies* Kyiv: RVV IMP B **18(1)** pp 107 –124
[11] Mandal P K. 2002 Anion water in gypsum (CaSO42H2O) and hemihydrate (CaSO4_1/2H2O) *Cement and Concrete Research* **32(2)** T K Mandal pp 313–316

[12] Mayeva I 2009 Structuring of Anhydrite Matrix with Nanodisperse Modifying Additives *Construction Materials Moscow* **6** pp 4–5
[13] Babuschkin W, Matwejew G and Mtschedlow-Petrosjan O 1986 *Thermodynamics of Silicates* Moscow: *Strojizdat* p 407
[14] Fischer H, Schlenkina S and Garkawi M 2000 Researching the process of aging of gypsum binders *Building materials and products Magnitogorsk* p 43–50

Physical and mechanical properties of hybrid-reinforced decorative concrete

A D Dovgan[1], V M Vyrovoy, S V Sukhanova

Odessa State Academy of Civil Engineering and Architecture,
Didrikhsona Str. 4, 65029 Odessa, Ukraine

[1]E-mail: aleks.dovhan@gmail.com

Abstract. The article presents the results of studies of the influence of alkali-resistant glass fibers of different lengths on the functional properties of decorative concrete. It is based on the data of an experiment carried out according to the optimal unsaturated plan, which includes 27 different compositions of the material. The combinations of the decorative compositions had equal mobility of the mixture, with the minimum and maximum water-cement ratio of 0.36 and 0.55, respectively. Based on the results of the experiment, the primary experimental-statistical models describing the effect of composition factors on the properties of the material made it possible to reveal that an increase in the level of plasticization of the compositions allows to increase the utilization factor of the strength of glass fibers. The efficiency of hybrid reinforcement of decorative composites with glass monofilaments has been confirmed through the implementation of a number of computational experiments (CE). The analysis of the CE results showed that the introduction of hybrid highly dispersed fibers into the initial composition of polystructural decorative concrete, with their different quantitative combinations, makes it possible to improve a number of its functional characteristics and thereby ensure the preservation of the artistic and aesthetic expressiveness of products.

1. Introduction

In recent decades, the construction industry has seen a significant demand for decorative building materials. This is due to the fact that elements and products made from such materials can significantly improve the architectural expressiveness of buildings and structures, the landscape design of adjoining and garden and park areas, while creating a unique variety of the environment [1]. In particular, the use of high-tech decorative materials contributes to obtaining architectural details of various shapes with a high level of detailing and an improved design of the front surface of the material in a broad sense (color, texture, texture, etc.).

However, the effectiveness of using decorative building materials is largely determined by the ability of decorative elements and products to maintain the specified functional properties during the normalized period of operation. In this context, functional properties are understood as a complex of physical and mechanical properties that ensure the preservation of artistic and aesthetic expressiveness under conditions of continuous changes in temperature and humidity under continuous exposure of gaseous and liquid chemically active substances. Such external influences lead to the emergence and development of alternating thermal and moisture deformations, which, taking into account the aggressiveness of the external environment, provoke a decrease in functional properties. One of the effective methods that cause a redistribution and change in the kinetics of the development of intrinsic and forced volumetric deformations is the use of highly dispersed reinforcement.

2. Analysis of scientific research, the purpose of the work

Decorative concrete, as a polystructural composite material, at all levels of structural inhomogeneities is characterized by the presence of a system of defects of different sizes (cracks, pores, voids, etc.). In the works of A.M. Brant, V.M. Vyrovoy, D.N. Korotkih and others [2-4] it is noted that the consequence of the defectiveness of composites is the localization of deformations and stresses with their concentration at the crack tips. It is known that cracks in the structure of cement composites spread due to the uneven distribution of volumetric deformations and stresses arising even under small internal and external influences, and due to a local excess of the tensile strength of concrete. In this regard, a rational solution to counteract and prevent the propagation of cracks of different sizes is the reinforcement of decorative concrete with hybrid discrete fibers at the micro and macro levels of its structure.

Analysis of scientific and technical literature shows that the type, quantity, geometric parameters and properties of hybrid discrete fibers, their orientation in relation to cracks, affect the change in the properties of building materials, products and structures. In particular, in the production of dispersed-reinforced decorative concretes, for which a high artistic and aesthetic expressiveness of the front surface of products is important, the basic principles of their technology should be taken into account [5]. For example, the type of concrete matrix, the type of dispersed reinforcement, methods of production of architectural products, giving the necessary texture to the front surface of products, etc.

The results of the researchers show that to control larger cracks, discrete fibers of greater length and high mechanical properties should be used, and to control small cracks - flexible and, accordingly, shorter lengths [6]. However, the research of scientists on hybrid concrete reinforcement at different levels of its structure is somewhat different. Thus, the article [7] shows the effect of hybrid steel fibers 50 and 25 mm long, taken in different proportions, on the mechanical properties of concrete. Mohammadi and others noted that the best effect on strength gain is obtained with a long to short fiber ratio of 65 + 35 %, respectively. In turn, the results presented in [8] on the hybridization of concrete with steel and polypropylene fibers demonstrate that steel fibers make the main contribution to the increase in strength, while polymer fibers do not have a significant effect. However, SU Jie and CHI Yin found a synergistic effect between the fibers in terms of deformability and resistance to external influences of the material. However, the experimental-statistical models (ES-model) obtained in our own research on the effect of a system of glass and polypropylene fibers on the strength and crack resistance of fine-grained concrete [9] made it possible to detect a negative synergy of two types of fibers. Therefore, the combination of high-modulus fibers of different lengths, but of the same type, can achieve the desired effect in strengthening the cement matrix of concrete. With regard to the fiber material, the most effective in improving the functional performance of decorative concrete are highly dispersed glass fibers of various lengths. At the same time, the tensile strength of glass monofilaments is slightly higher compared to the strength of hardened concrete. In particular, analysis of numerous works shows the advisability of reinforcing concrete at the micro level with fibers $10 \div 30$ μm in diameter and less than 10 mm in length, and, accordingly, at the macroscale level – $10 \div 60$ mm long.

Due to the fact that discrete reinforcing fibers are the initial components, it can be assumed that they are actively involved in organizing the structure of the material and through structural transformations affect the change in the properties of the material itself, as well as products, individual structural and decorative elements and structures. This predetermined the goal of research – increasing the functional properties of decorative items through the targeted use of the effects of multi-site structure formation by introducing hybrid dispersed fibers into the initial composition of polystructural materials.

3. Experimental conditions, functional properties of decorative concrete

Based on the conclusions of materials scientists in the field of research of fine-grained composites and regulatory requirements for the quality criteria of architectural details, when developing compositions of decorative concrete, recipe factors were selected and the ranges of their variation were determined [10].

The experiment was carried out according to the optimal symmetric plan, which includes 27 different compositions. Varying at three dosage levels of 5 components. Depending on the nature of their action on a fine-grained composite, the factors are combined into two groups.

The conditions for modifying the cement-sand system set the amount of finely ground zeolite X_1 (Z) at levels of 0, 4, 8 % (introduced instead of part of the cement from its mass), the proportion of fine quartz sand mixed with coarse X_2 (GS) – 30, 50, 70 % and a polycarboxylate-based superplasticizer from the Melflux series X_3 (MF) – 0.3, 0.5, 0.7% (based on the mass of cement).

The group of dispersed reinforcement factors consisted of alkali-resistant glass fibers with a length of 6 and 12 mm at the levels X_4 ($F6$) $= X_5$ ($F12$) – 0, 0.015, 0.03 % (were introduced from the mass of the mixture).

The varied factors are normalized to the condition $-1 \le x_i \le +1$ by typical formulas [11].

The fixed components of the decorative composite were the dosages of white Portland cement and quartz sand of natural granulometry, at a ratio of 1:3.4 by mass of the mixture. The compositions of the mortar compositions had equal mobility of the mixture (grade $S4$), with the minimum and maximum water-cement ratio of 0.36 and 0.55, respectively.

A number of functional indicators of decorative concrete were determined, in particular, the compressive strength (f_{cm}, MPa) and tensile strength in bending (f_{ctfm}, MPa) in equilibrium (index "n"), dry ("d") and water-saturated ("w") states of material, crack- (K_{1c}, MPa·m$^{0.5}$) and frost resistance (F, cycles), water absorption by weight (W_M, %). At the same time, to study the indicated properties of the material, in accordance with the experimental plan, samples were made and tested in accordance with the requirements of a number of regulatory documents.

According to the experimental data, a complex of ES-models of the form (1) was built, which describe the fields of functional characteristics (Y_i) of decorative concrete in the coordinates of 5 prescription factors $x_1 \div x_5$ [12-15].

$$Y_i = b_0 \boxed{\begin{array}{l} \pm b_1x_1 \;\pm b_{11}x_1^2 \;\pm b_{12}x_1x_2 \\ \pm b_2x_2 \;\pm b_{22}x_2^2 \;\pm b_{13}x_1x_3 \\ \pm b_3x_3 \;\pm b_{33}x_3^2 \;\pm b_{23}x_2x_3 \quad \textbf{(a)} \end{array}} \boxed{\begin{array}{l} \pm b_{14}x_1x_4 \;\pm b_{15}x_1x_5 \\ \pm b_{24}x_2x_4 \;\pm b_{25}x_2x_5 \\ \pm b_{34}x_3x_4 \;\pm b_{35}x_3x_5 \quad \textbf{(c)} \end{array}}$$
$$\boxed{\begin{array}{l} \pm b_4x_4 \;\pm b_{44}x_4^2 \;\pm b_{45}x_4x_5 \\ \pm b_5x_5 \;\pm b_{55}x_5^2 \quad \textbf{(b)} \end{array}} \qquad (1)$$

Nonlinear ES-model (1) is structured by groups of factors. Block (a) describes the effect of modifiers x_1, x_2 and x_3 on the analyzed property Y of the composite. Block (b) – influence on Y of dispersed fibers x_4 and x_5. Block (c) shows how the local fields of the analyzed property will be transformed due to the relationship between the two groups of composition factors.

In particular, on the basis of mathematical models (1) on the kinetics of strength gain in compression of a composite on days 1, 3, 7, 14 and 28, a number of useful results were obtained on the complex effect of prescription factors on f_{cm} of decorative concrete. At the same time, it was revealed that an increase in the level of plasticization of the composites to the maximum (MF from 0.3 to 0.7%) makes it possible to increase the utilization factor of the strength of glass fibers. This can be explained by the fact that the cement matrix of highly plasticized composites is characterized by apparently lower porosity in the zone of contact with the fiber. Consequently, the importance of the quality of the cement matrix is confirmed when using smaller fibers of two types [2]. Also, as noted in [16], with a properly designed composition with fine ingredients, when the size of the filler particle and the fiber cross-section is less than or equal to the size of the cement grain, it is possible to achieve high adhesion of the matrix to the fiber and at the same time to obtain a denser microstructure of the material. It should be noted that in the design of decorative concrete compositions, the average particle size of cement and finely ground zeolite ($\varnothing_p \approx 15$ μm), as well as the cross section of glass fibers ($\varnothing_f \approx 14$ μm) were taken into account. Therefore, it seems appropriate to analyze the effect of highly dispersed hybrid fibers ($F6 + F12 \rightarrow 0.03 \div 0.06$ %) on the functional properties of concrete with the maximum level of plasticization ($MF = 0.7$ %).

4. Results of computer modeling of functional properties of dispersed-reinforced decorative concrete

To determine the effectiveness of hybrid reinforcement of decorative composites with glass fibers of different lengths (6 mm and 12 mm), a number of computational experiments (CE) were performed. For the implementation of CE, structured ES-models (1) [17] of the functional properties of concrete (f_{cm}, f_{ctfm}, K_{1c}, W_M) were used in the coordinates of normalized factors $x_1 \div x_5$. At the same time, for the characteristics of frost resistance, a "complete" model of compressive strength ($f_{cm.F}$, MPa), built on the basis of the test results of concrete specimens tested for the declared grade $F400$, was taken as a basis [14]. Since CE are carried out under the condition of high plasticization ($x_3 = +1$) of decorative composites and the content of hybrid glass fibers at different levels ($x_4 + x_5 \rightarrow 0$ and $+1$) from ES-models of the "full field" for each property [12-15] (Y_i) obtained 2-factor models of the form (2).

$$Y_i = b_0 \pm b_1x_1 \pm b_{11}x_1^2 \pm b_{12}x_1x_2 \pm b_2x_2 \pm b_{22}x_2^2 \quad (2)$$

Models (2) describe local fields Y_i (figure 1 and figure 2), formed by two factors x_1 $\{Z\}$ and x_2 $\{GS\}$, at $x_3 = \text{const}$, under the influence of dispersed reinforcement factors ($x_4 + x_5$). It should be noted that in order to compare the effect of hybrid reinforcement of a cement-sand system, a 2-factor model Y_i was also taken into account, obtained with fixed dosages of fibers at low levels ($x_4 = x_5 = -1$), i.e. in their absence. Thus, in the diagram of each Y_i property (figure 1 and figure 2), two surfaces are shown, which are formed by reinforced and unreinforced compounds of the composite. The change in the property Y_i within the boundaries of the local field was estimated by its absolute and relative increments [17].

The obtained results of modeling the physical and mechanical properties of the decorative composite allow us to note the following.

The maximum compressive strength f_{cm} is achieved exclusively in concretes on coarse sand ($x_2 = -1$ or $GS = 30$ %), reinforced with high and medium dosages of short and long fibers ($F6 + F12 = 0.045$ %). Moreover, to achieve the best effect, the introduction of finely dispersed zeolite ($x_1 = -1$ or $Z = 0$ %) is not required. At the same time, the strength of composites in equilibrium $f_{cm.n}$ and dry $f_{cm.d}$ states, in comparison with unreinforced material compositions, increases by 14 % and 24 % (figure 1a).

The flexural tensile strength f_{ctfm} of concretes is maximum when the content of hybrid fibers is at medium levels ($F6 + F12 = 0.03$ %). Under these simulation conditions, f_{ctfm} increases from 4.2 % to 7.9 % for composites in a water-saturated $f_{ctfm.w}$ and dry $f_{ctfm.d}$ state (figure 1b). However, to achieve this effect, finely ground pozzolan should be introduced instead of a part of the binder at least $Z \geq 4$ % (or $x_1 \geq 0$).

The introduction of hybrid fibers (at different ratios) into highly plasticized composites makes it possible to increase their crack and frost resistance by more than 6 % and 2.6 %, respectively. The increase in crack resistance is confirmed by the analysis of the data of the coefficient of technological influence $K_T = K_{1c}^L / K_{1c}^S$ (proposed by V.N. Vyrovoy, [18]), and the ability of concrete resistance to frost fracture - the compressive strength $f_{cm.F}$ of the material samples that have passed the test (figure 1c and 1d).

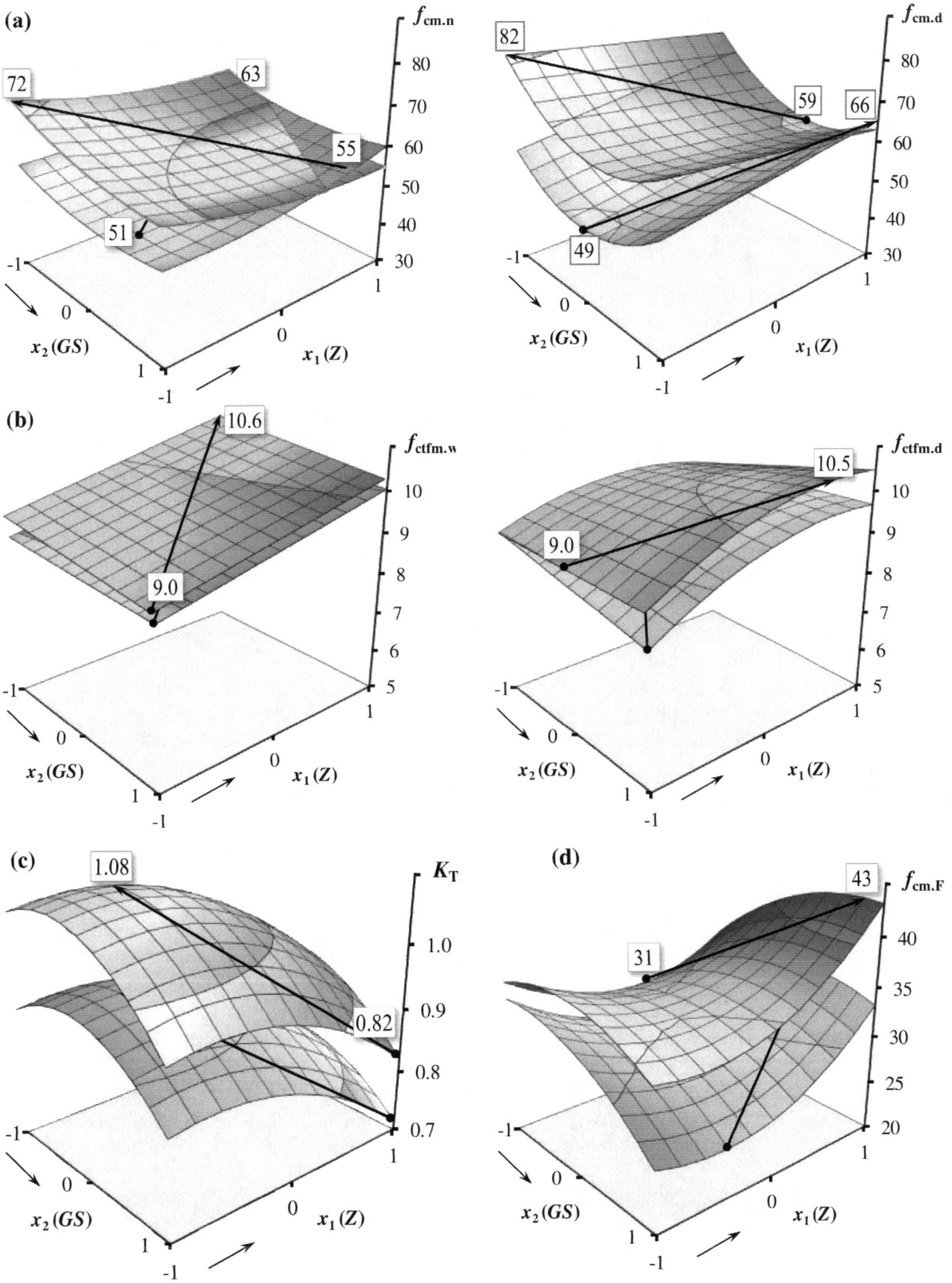

Figure 1. Local fields of properties of dispersed-reinforced (top surface) and unreinforced (bottom surface) decorative concrete: **(a)** f_{cm}, **(b)** f_{ctfm}, **(c)** K_T and **(d)** $f_{cm.F}$

The positive effect of high-modulus and high-strength fiberglass is explained by the fact that they are capable, by means of tangential forces acting on the interface, to perceive the bulk of the applied stresses and redistribute volumetric deformations over the entire cross-section of the material. As a result, the cracking process is inhibited and the physical and mechanical characteristics of concrete are improved (Rabinovich F.N., Pashchenko A.A., Ramachandran V., Biryukovitch K.L. *et al.*) [19-20 *et al*].

In addition, positive dispersed reinforcement of concrete is observed in the study of water absorption at hydrostatic pressure. The content in the material of medium and high dosages of short and long fibers (0.9 kg per 1 m^3 of mixture) does not increase the saturation of concrete with water, and even reduces it by 10 % (figure 2). However, this result is achieved when plasticizing the

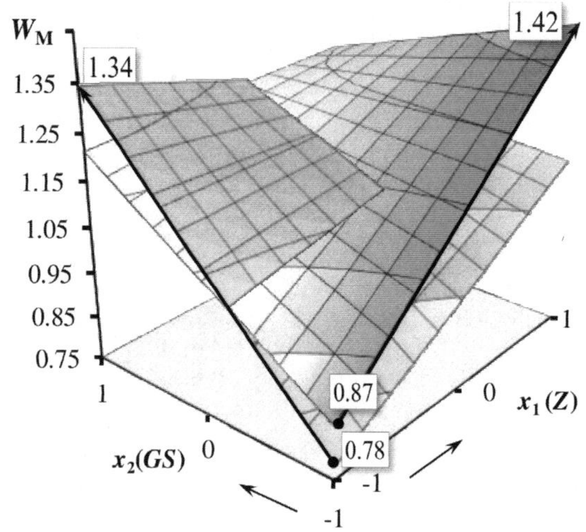

Figure 2. Local water absorption of dispersed-reinforced (max = 1.34 % and min = 0.78 %) and unreinforced (max = 1.42 % and min = 0.87 %)

compositions by adding polycarboxylate at the level of MF = 0.54 % (or x_3 = +0.22). Therefore, it can be concluded that the microfibers are efficiently packed between the cement grains. It is important to note that according to the fiber manufacturer, 1 kg contains approximately 200 million fibers.

5. Conclusion

The experimental-statistical "complete models" constructed on the basis of the results of the experiment, describing the influence of the composition factors on the properties of the material, revealed that an increase in the level of plasticization of the compositions rises the utilization factor of the strength of glass fibers. The effectiveness of hybrid reinforcement of decorative composites with glass monofilaments has been confirmed through the implementation of a number of computational experiments. The analysis of the experimental results showed that the introduction of hybrid highly dispersed fibers into the initial composition of polystructural decorative concrete, with their different quantitative combinations, makes it possible to improve a number of its functional characteristics and thereby to ensure the preservation of the artistic and aesthetic expressiveness of the product.

References

[1] Architectural Precast Concrete *PCI Architectural Precast Concrete Manual Committee* 2007 (Sidney Freedman Editor et al.) p 588
[2] Andrzej M Brant 2009 *Cement-based composites: materials, mechanical properties, and performance* (Taylor & Francis, London and New York) p 536
[3] Vyrovoy V, Sukhanov V and Korobko O 2016 *Material structure in structure of construction* (Odessa: Poligraf) p 244
[4] Korotkih D 2011 *Stroitel'nye Materialy [Construction Materials]* No 3 pp 96-99
[5] Dovgan P 2021 *Dispersed-reinforced decorative concretes for architectural elements* Manuscript (Odessa: Poligraf) p 196
[6] Vahid Afroughsabet1, Luigi Biolzi1, and Togay 2016 *High-performance fiber-reinforced concrete: a review* (Springer Science+Business Media New York)
[7] Mohammadi Y, Singh SP, Kaushik SK 2008 *Properties of steel fibrous concrete containing mixed fibres in fresh and hardened state* Constr Build Mater 22(5):956-965
[8] SU Jie, CHI Yin 2015 *Stress-strain Behavior of Steel-Polypropylene Hybrid Fiber Reinforced Concrete Under True Triaxial Compression.* 5th International Conference on Civil Engineering and Transportation (ICCET 2015) pp 735-742
[9] Lyashenko T V and Dovgan A D 2016 Mechanical properties of fine-grained concrete with microsilica and glass fibre *Bulletin of Odessa State Academy of Civil Engineering and Architecture* (OGASA Odessa vol 62) pp 110-114.
[10] Dovgan P M, Dovgan A D, Lyashenko T V and Hlitsov N V 2018 Experiment planning in the study of decorative sand concrete *Structure formation, strength and destruction of composite building materials and structures* (OGASA Odessa) pp 27-30
[11] Voznesensky V A, Lyashenko T V and Ogarkov B L 1989 *Numerical methods of solving building technological problems with computer* (Kiev: Vischa Shkola) p 328
[12] Lyashenko T V, Dovgan A D and Dovgan P M 2018 Fibre-reinforced decorative composite: the effects of composition on the strength *Bulletin of Odessa State Academy of Civil Engineering and Architecture* (OGASA Odessa vol 73) pp 89-97
[13] Dovgan A D, Vyrovoy V M and Dovgan P M 2019 Crack resistance of decorative composites IOP *Conf. Series: Materials Science and Engineering* vol 708
[14] Dovgan A D, Vyrovoy V M, Dovgan P M and Silchenko S V 2020 Frost resistance of decorative composites of different composition *Actual Problems of Engineering Mechanics: Materials Science and Technologies* pp 35-42
[15] Dovgan A D, Vyrovoy V M and Dovgan P M 2020 Water absorption as a function of decorative concrete warehouses *Collection of scientific works, Resource and economic materials, structures, buildings and constructions* (NUWEE Rivne no 38) pp 148-160
[16] Arnon Bentur, Sidney Mindess 2007 *Fibre reinforced cementitious composites*. 2nd ed. *Modern Concrete Technology Series* (Taylor & Francis, London and New York) p 601
[17] Lyashenko T and Voznesensky V 2017 *Composition-process fields methodology in computational building materials science* (Odessa: Astroprint) p 168
[18] Solomatov V I, Vyrovoy V N, Dorofeev V S, Sirenko A V 1991 *Composite building materials and constructions of the reduced material capacity* (Kiev, Budivelnik) p 144
[19] Rabinovich F N 2004 Composites based on dispersed reinforced concrete *Questions of theory and design, technology, designs: monograph* (Moscow, ASV Publishing House) p 560
[20] Pashchenko A A 1979 The reinforcement of inorganic binders with mineral fibers (Kiev, Vischa Shkola) p 200
[21] Ramachandran V, Feldman R and Beaudoin J 1981 *Concrete science: Treatise on Current Research* Heyden and Son, Ltd (London – Philadelphia – Rheine) p 278
[22] Biryukovitch K L, Biryukovitch Y L and Biryukovitch D L 1986 Glass cement in construction (Kiev: Budivelnik) p 96

Features of the organic-mineral intumescent paints structure formation for wooden constructions fire protection

S Guzii[1], T Kurska[2], Y Otrosh[3], P Balduk[4], Y Ivanov[5]

[1,2,3,5]National University of Civil Defense of Ukraine, 94, Chernyshevska st., Kharkiv 61023, Ukraine
[4]Odessa State Academy of Civil Engineering and Architecture, 4, Didrikhsona st., Odessa, Ukraine, 65029
[3] Address all correspondence to this author. E-mail: tkurskaya67@gmail.com

Abstract. The paper presents the results describing the features of the organic-mineral paints structure formation of the intumescent type for wooden structures fire protection. On the research basis a diagram of the coke formation process has been developed. It is shown that the introduction of titanium oxide and aluminum hydroxide in an amount of 5% promotes the ordering of the foam layer structure, an increase in its thermal stability and the achievement of the highest value of the swelling coefficient – up to 47, which is 1.7 times higher than the values of the swelling coefficient of the base composition. It was found that at 700 °C the formation of refractory compounds of the type TiP_2O_7, $Mg_3(PO_4)_2$, $AlPO_4$ occurs, which prevent the burnout of the formed coke foam and increase its strength and heat resistance. On the basis of the main provisions of fuzzy mathematics and data from physicochemical studies, a model of the mechanism of structure and coke formation and control of this process is proposed.
Keywords: organic-mineral paints, swelling, fire protection, wooden constructions, structure formation, foam coke, mineral components, coke-formers.

1. Introduction and literature review

During 2020 fires destroyed and damaged 28029 buildings and structures of various purposes, of which 7106 were completely destroyed, including 796 residential buildings. Of the total number of buildings and structures destroyed and damaged by fires in 2020 – 11707 (41.8%) – had III degree of fire resistance, 5373 (19.2%) – V degree of fire resistance, 2889 (10.38%) – had IIIa degree of fire resistance. Of the 7106 buildings completely destroyed in 2020, 2263 (31.9% of the total number of destroyed buildings and structures) had a V degree of fire resistance, 1531 (21.6%) had a III degree of fire resistance, 841 (11.8%) – IIIa degree of fire resistance, 431 (6.1%) – IV degree of fire resistance, 246 (3.5%) – IIIb degree of fire resistance, 108 (1.5%) - IVa degree of fire resistance, 86 (1.2%) – II degree of fire resistance and 12 (0.2%) – I degree of fire resistance. Wooden constructions are used in buildings and structures of III, IIIb, IV degrees of fire resistance with mandatory fire protection [1].

Recently in Ukraine there has been an increased interest in the results of scientific developments in the field of creating highly effective means of protecting building structures from the effects of fire and their implementation. [2].

One of the factors on which fire safety is based during the design and operation of buildings and structures for various purposes is to ensure fire resistance of building structures and their ability to spread fire. Compliance with the principles of fire safety in construction is based on an engineering approach to assess the required level of safety in the design, construction, expansion, reconstruction, technical re-equipment, overhaul, change the functional purpose of buildings and structures necessary measures and safety tools. The basic requirement for fire safety in construction is to limit the spread of fire, smoke and maintain the load-bearing capacity of the structure for a certain period of time, and an important way to ensure these principles is to manufacture fire-resistant building structures and study their fire resistance, including fire-resistant constructions.

In this regard, the need to research in this direction is determined with the provision of special attention to the development of effective fire retardant coatings for the purpose of their use in the construction of both general construction and special purpose facilities, where the use of impregnating mixtures is ineffective [3]. In addition, a number of resonant accidents at military facilities (Artemovsk, 2003, Novobogdanovka, 2004, Tsvitokha, 2005, Lozovaya, 2008) revealed a low level of safety assurance for the operation of arsenals, bases and warehouses in connection with the use of wood in building constructions, structures and storage facilities that are highly sensitive to the effects of high temperature and fire, and therefore, to improve the safety of operation of ammunition storage facilities, to prevent and limit the hazardous factors of combustion of wood products, fire protection of materials is used [4-6]. Analyzing the results on the development of fire retardant coatings based on alkaline aluminosilicate binders [7, 8], it was established that a significant thickness of the protective layer should be applied to building structures and the coatings become rigid during operation (low elasticity), which leads to a loss of adhesion properties and shedding under the influence of temperature and humidity fluctuations [9-11].

Modern fire retardant coatings are characterized by the formation of an expanded layer of coke on the surface of a building construction [12-14], which significantly reduces the processes of heat transfer to the material, and are complex systems of organic and inorganic components [15-17], but when exposed to flame they are characterized by high smoke-forming ability and toxicity [18-20].

As an alternative solution to the problem of wooden constructions fire protection, it is possible to use coatings based on organic-mineral binders, which will reduce the above indicators by adding mineral fillers to the organic base and form a more stable expanded foam coke layer [21, 22].

As shown by the analysis of scientific research [23, 24], in the formation of the predicted level of quality, in particular, the functionality and durability of coatings, an important place is occupied by the type of film-former of the original composition, the type of filler and the structure of the coating. However, they cannot provide reliable protection of constructions at temperatures above 1273 K, since at higher operating temperatures, the organic component is destroyed, and the coating becomes porous, which significantly worsens its operational properties.

2. Literature review

Foaming and coke formation of intumescent coatings is accompanied by various physicochemical processes, which, as a rule, proceed in a certain sequence as the temperature effect on the composition increases [25-28].

The swelling mechanism of such coatings has not been sufficiently studied. This is due to the fact that the main reactions leading to the production of a protective foam-coke layer occur in the high temperature range (up to 900 °C), which complicates the modeling of these processes. In addition, they belong to multicomponent composite materials. This leads in turn to a large number of possible interactions between the components formed by the fire retardant coating, especially at high temperatures. At the same time, it is also rather difficult to foresee the direction of high-temperature reactions. According to many experts, the swelling mechanism of fire retardant coatings consists of the following stages [29]:

- an inorganic acid evolution capable of esterification of carbonization of materials (polyols) at 150-215 °C, depending on the type of acid derivative;
- the esterification of the polyol was acidified at a temperature just above the temperature of the first stage. In the presence of amines, the reaction is accelerated and completed at a lower temperature;
- melting the mixture of intumescent coating components immediately prior to or during esterification;

- chemical transformations - the formation of an ester of a polyol with an inorganic acid as a result of dehydration, resulting in the formation of carbon-inorganic (usually carbon and phosphorus) structures, which are further involved in coking processes;
- foaming the carbonaceous mass with incombustible gases that are released during the decomposition of the foaming agent, as well as water vapor that is formed as a result of dehydration of the polyol;
- gelatinization, and then solidification of the foamed mass at the end of the coking process.

This solid mass consists of finely porous foam and is a good heat insulator. An example of a different interpretation of the mechanism of swelling of coatings can be the data of [30]. In particular, based on the results of a comprehensive thermal analysis of various mixtures of components of organic-mineral coatings and IR spectroscopy of coke foam, it is assumed that pentaerythritol is not esterified with phosphoric acid, but decomposes under conditions of elevated temperature and acidic environment into formaldehyde and acetaldehyde (Figure 1):

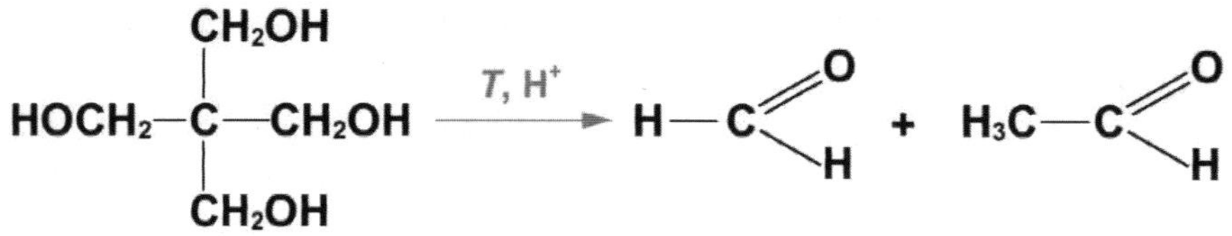

Figure 1. Reaction of pentaerythritol decomposition during coating swelling.

Aldehydes formed as a result of the decomposition of pentaerythritol are involved in the synthesis of aldehyde-melamine resin, a precursor of the foam-coke framework (Figure 2).

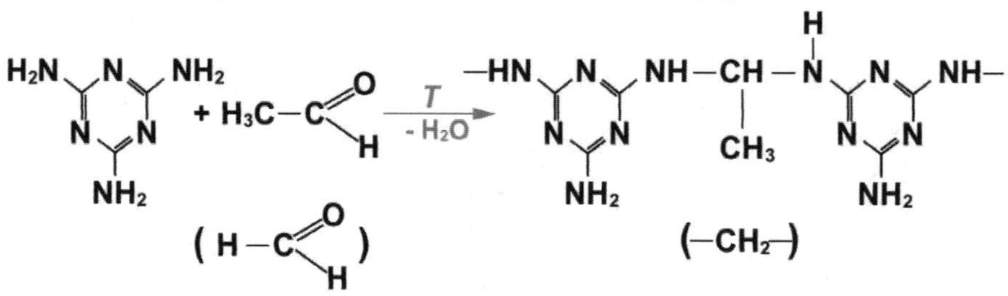

Figure 2. A foam-coke framework formation.

In turn, ammonium polyphosphate acts as a growth regulator for the aldehyde-melamine resin, being chemically bonded on one side to the substrate, and on the other to melamine.

Typically, a fire retardant coating contains the following main components: a catalyst that decomposes when the temperature rises to form an inorganic acid; a carbohydrate, which reacts with said acid to form a carbon layer; a binder that melts when the temperature rises, contributes to the formation of a charred insulating layer; swelling agent, which decomposes simultaneously with the melting of the binder, releases a significant amount of non-combustible gases and is accompanied by foaming and multiple (up to 40 times) increase in the thickness of the coating, transformed into an insulating carbon layer that protects the building structure from further exposure to high temperatures [31].

Fire retardant intumescent compositions are rather complex multicomponent systems, since they include three main components: a coke-formation catalyst, coke-forming and foaming agents. As a rule, the catalyst is phosphorus-containing compounds and, most often, ammonium polyphosphate. The content of ammonium polyphosphate in the composition in many formulations is explained by its participation in the formation of the coke structure. As a raw material for the formation of the carbon framework of the foamed layer, polyalcohols are usually used, and as a porophore, organic amines or amides, which emit non-combustible gases at elevated temperatures – carbon dioxide, nitrogen, ammonia, which foam the system. The most popular in modern formulations of intumescent coatings from polyalcohols are glycerin and pentaerythritol, and from amines – melamine and carbamide. To perform conventional protective and decorative functions and for long-term storage of fire retardant characteristics during operation, thermoplastic polymers are most often used as film formers in the manufacture of fire retardant intumescent compositions, namely vinyl acetate homo- or copolymers or other water-dispersive binders [32, 33].

In the initial state, fire retardant compositions are solid heterogeneous, heterophase granular systems, consisting of randomly spaced particles ranging in size from 1 to 40 microns and a large number of pores. When the fire retardant coating is heated, a series of chemical reactions begins: ammonium polyphosphate decomposes and releases phosphoric acid, phosphoric acid causes pentaerythritol or dipentaerythritol to dehydrate to form a protective layer for such materials.

The use of minerals in organic-mineral compositions can reduce smoke generation and increase weather resistance and develop a new type of fire retardant coatings for building constructions, the presence of which will slow down the heating of the material and maintain their functions during a fire for a given period of time.

According to [21, 26, 27, 34] fillers as formers of a porous structure are divided into two groups:

- contribute to the formation of a porous (foamed) structure of coke (titanium dioxide, borates of barium and zinc, magnesium hydroxide)
- suppress the formation of foam in coke (aluminum hydroxide, intumescent graphite, zeolite, sodium tripolyphosphate).

Compounds of the first group act as active bubble generators and cause the formation of a more regular and stable foam in comparison with foam-coke produced from similar compositions without filler (pigment). The most effective foaming agents are titanium dioxide and magnesium hydroxide.

In turn, the second group compounds suppress the formation of foam in coking layer. For example, intumescent graphite provides coke with its structure, while suppressing foaming of the organophosphate mass, and foamed coke, which is formed in the presence of sodium tripolyphosphate, has significant fluidity.

Inorganic fillers contribute to the swelling of coatings, which decompose at elevated temperatures with the formation of non-combustible gaseous products (for example, CO_2, water vapor) and remove heat. These include aluminum hydroxide, calcium, cadmium and zinc carbonates, boric acid salts.

Materials such as zinc borate, siloxanes, talc, kaolin, fiberglass and others are added to the compositions of organic-mineral paints of the intumescent type to form glassy and ceramic structures in coke at elevated temperatures, and strengthen the foam-coke framework. For example, at temperatures up to 600 °C, ammonium polyphosphate interacts with talc to form crystalline products such as $Mg_2P_4O_{12}$, MgP_4O_{11}, SiP_2O_7, $Si_5O(PO_4)_6$, and at temperatures close to 1000 °C, to form glassy magnesium and silicon phosphates.

Titanium dioxide at a temperature of about 600 ° C reacts with ammonium polyphosphate to form titanium pyrophosphate TiP_2O_7 – a fireproof material that stabilizes the insulating foam at high temperatures, when most of the carbon is oxidized and burned out to form CO_2. In this case, a white bloom appears on the coke layer [28, 29, 35].

Intumescent fire retardant paints/coatings are widely used and this is due to their low thermal conductivity under fire conditions due to the formation of a fine-meshed carbon layer of the coating, which complicates the heating of wood, lengthening the phase of its preparation for active participation in the combustion process. During swelling, the components soften with simultaneous endothermic decomposition of fire retardants and blowing agents, which increases the fire-retardant properties of coatings during swelling [25, 32].

The aim of this work is to study the processes of structure formation of organic-mineral paints of the intumescent type for fire protection of wooden constructions. The implementation of the goal of the work is carried out by solving the following tasks, namely: to develop a scheme for the process of coke formation; to study the influence of mineral fillers on the processes of structure formation; to develop a mathematical model of the processes of structure formation of organic-mineral paints of the intumescent type.

3. Materials of research

To obtain organic-mineral paints of the intumescent type, polyvinyl acetate (PVA) dispersions were used as a binder [36]; as a catalyst and coke-forming agent – ammonium polyphosphate (PFA) [37], melamine [38], pentaerythritol (PER) [39] (table 1); as mineral fillers – titanium dioxide, talc, aluminum and magnesium hydroxide, and their mixtures [40].

Table 1. Characteristics of organic components

Component	Chemical formula	Melting temperature, °C	Solubility in water at 25 °C, g/100 ml	Density, g/sm^3	Particle size, microns
PFA	$(NH_4PO_3)_n$	312	0.8	1.9	15
M	$C_3H_6N_6$	354	3.2	1.573	40
P	$C(CH_2OH)_4$	215	7.1	1.394	40

Note: PFA – ammonium polyphosphate; P – pentaerythritol; M – melamine

Organic-mineral paint of the intumescent type was obtained by the technology of water-dispersion paints with the introduction of the necessary technological additives in a laboratory dissolver [40].

4. Methods of research

The study of features of structure formation processes was performed using X-Ray diffraction (XRD), differential thermal analysis (DTA) and digital microscopy.

X-ray phase analysis was carried out on a DRON-3M and DRON-4-07 diffractometer with a copper tube at a voltage of 30 kV, a current of 10 ... 20 mA, and a range of angles 2Θ=10...60° at a counter rotation speed of 2° per minute. New materials were identified based on the PDF-2 Data Base (Sets 1-50 plus 70-88) with the JCPDFWIN 2.1 software module (JCPDS-ICDD, 2000). Differential thermal and thermogravimetric analysis was carried out on a derivatograph of the R. Paulik, I. Paulik, L. Erdey by the MOM company (Hungary). The samples were heated at a rate of 10 °C per minute to a temperature of 1000 °C.

The obtained experimental data were interpreted according to [41, 42]. The macrostructure of the foamed paint layer was determined using a Dino-Lite Pro-AM413T5 digital microscope manufactured by ANMO Electronics Corporation (Taiwan) with a 1.3 Mpix camera at x500 digital magnification.

Determination of the coefficient of swelling of fire-retardant coatings was determined according to [43].

5. Study of mineral composition, size particles and surface area of quartz dispersed materials

It is possible to increase the thermal and mechanical stability of the coke foam to the effects of variable heat fluxes, besides modifying the organic component, by introducing refractory oxides and reactive water-containing reagents into the base composition, which at high temperatures pass into anhydrous phases, and contribute to an increase in its thermal stability.

Considering that an aqueous solution of modified PVA dispersion was chosen as a binder in the studies, it is hypothetically possible to imagine the mechanism of dissolution of ammonium polyphosphate in it, the interaction of the products of its dissociation with fillers: $Al(OH)_3$, TiO_2, $Mg(OH)_2$ and $Mg_3[Si_4O_{10}](OH)_2$:

I partial dissociation of ammonium polyphosphate and pentaerythritol with the formation of:

$Al_{3-6}[PO_4]_{2-4}[OH]_{3-6}\cdot(5\text{-}6)H_2O$
$Al_2(PO_4)(OH)_3$
$(NH_4)_m(HPO_4)_n + H_2O + Al(OH)_3 \rightarrow$
$AlPO_4\cdot 4H_2O$
$Al_4(PO_4)_3(OH)_3\cdot 9H_2O$
$Ti(OH)_2(HPO_4)\cdot 2H_2O$
$(NH_4)_m(HPO_4)_n + H_2O + TiO_2 \rightarrow$
$Ti(HPO_4)_2\cdot H_2O$
$(NH_4)_m(HPO_4)_n + H_2O + Mg(OH)_2$
$\rightarrow Mg_5Al_{12}(PO_4)_8(OH)_{22}\cdot 32H_2O$
$(NH_4)_m(HPO_4)_n + H_2O + Mg_3[Si_4O_{10}](OH)_2$

II partial decomposition of ammonium polyphosphate, pentaerythritol and phosphates Al, Ti and Mg with the formation of phosphate clusters:

$T \approx 250 - 280\ ^oC$

$(NH_4)_m(HPO_4)_n \rightarrow {}_mNH_3 + (HPO_3)_n$

III formation of coke foam due to decomposition, foaming, dehydration and reactions in solid phases:

$T \approx 300 - 1000\ ^oC$

$(HPO_3)n$, pentaerythritol, melamine, PVA dispersion, phosphate Al, Ti and Mg, dehydration of $Mg_3[Si_4O_{10}](OH)_2$ with the formation of anhydrous mineral phases capable of reinforcing the foam-coke layer and increasing its thermal stability and mechanical strength.

Based on the scheme of the coke formation process, in the base composition, according to XRD data (Figure 3, curve 1), diffraction bursts of type II ammonium polyphosphate insoluble in water are recorded – $(NH_4)m(HPO_4)n$ d = 0.939; 0.781; 0.696; 0.605; 0.566; 0.557; 0.534; 0.482; 0.486; 0.457; 0.4221; 0.389; 0.374; 0.349; 0.312; 0.307; 0.265; 0.260; 0.251; 0.248; 0.237; 0.223; 0.211; 0.201; 0.187; 0.176; 0.172; 0.168 0.160; 0.155; 0.152 nm. There are no diffraction bursts of pentaerythritol and melamine in the diffractogram. Their presence is confirmed by DTA data on a number of endothermic effects characteristic of these components (Figure 4, curve 1). This curve shows the exothermic effect in the temperature range (+) 60–70 °C, which explains the partial oxidation (resinification) of pentaerythritol. With an increase in temperature, partial destruction of pentaerythritol occurs, as evidenced by the presence of endothermic effects on the DTA curve at temperatures: (–) 255, 300 and 350 °C, which describe the chemical transformations of this component of the binder, namely, the passage of the pyrolysis stage with the formation of methanol, formaldehyde, the release of water and subsequent destruction in the direction of the formation of acrolene, carbon dioxide, resin residue and the synthesis of alphamethylacrolene.

Simultaneously with this process, the destruction of the PVA dispersion occurs - the endothermic effect at a temperature of (–) 200 °C, which is associated with the release of acetic acid and the formation of double bonds in carbon atoms. An increase in the action temperature leads to the thermooxidation of the PVA dispersion - an exothermic effect at a temperature of (+) 400 °C and the thermal decomposition of melamine, which is characterized by significant gas release, which leads to the formation of a porous layer of coke foam of a disordered structure (Figure 3, pos. A).

The process of coke formation is similar for other compositions, but something with its own specifics. Thus, the introduction into the organic-inorganic composition of 10% $Al(OH)_3$ d = 0.309; 0.248; 0.229; 0.204; 0.169; 0.162; 0.145 nm (Figure 3, curve 2) leads to an increase in the swelling coefficient (Table 2) due to two-stage dehydration of hydroargilite: endoeffects in the temperature range (–) 250–300 and 500–550 °C (Figure 4, curve 2). A further increase in temperature contributes to the additional release of water vapor from the hydroxide structure and contributes to the polymorphic transition of $\gamma\text{-}Al_2O_3$ to $\alpha\text{-}Al_2O_3$ exothermic effect at a temperature of (+) 800 °C and the

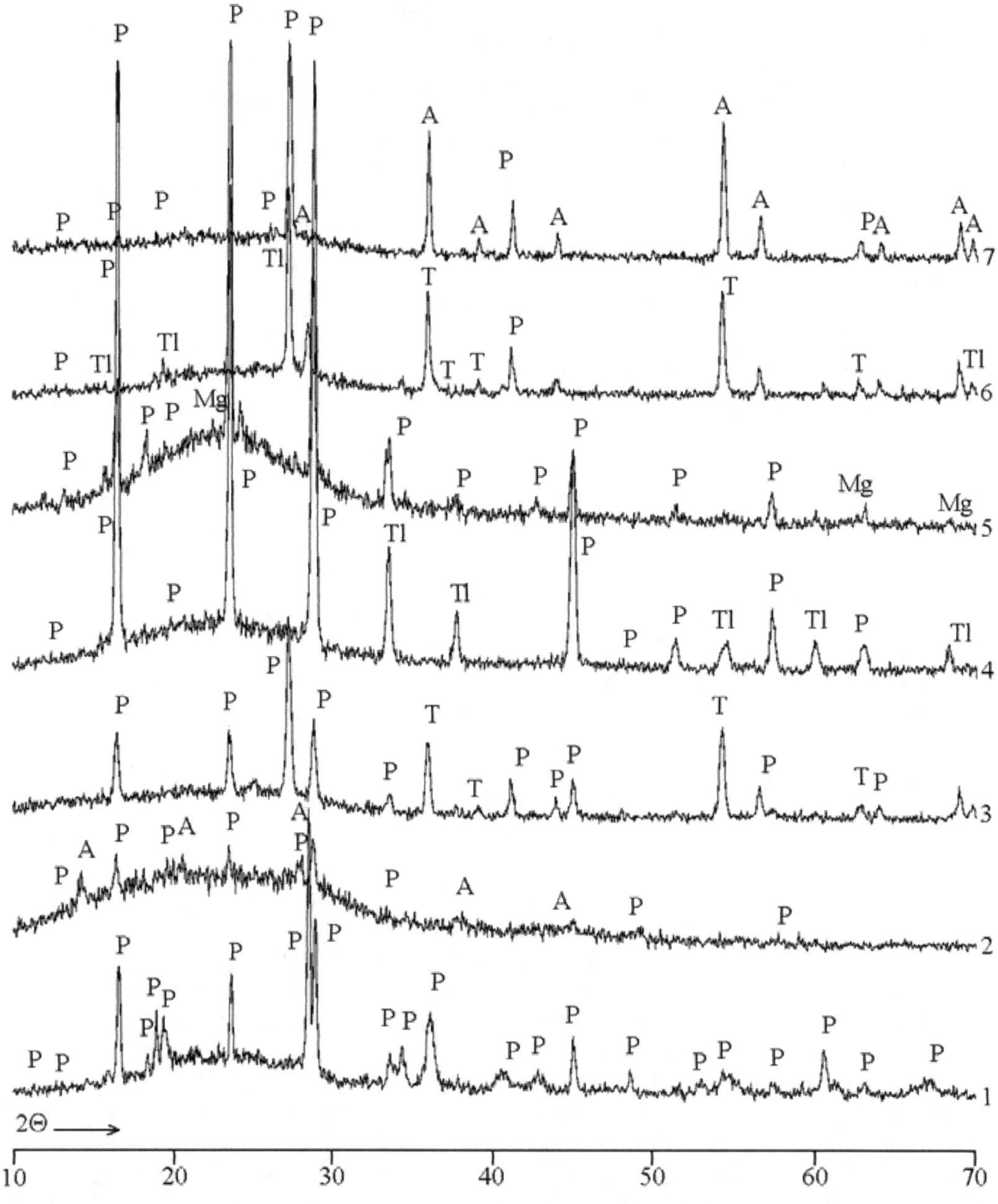

Figure 3. Sciagrams of an organic-mineral fire retardant composition: 1 – base composition; 2 – base + 10% $Al(OH)_3$; 3 – base + 10% TiO_2; 4 – base + 10% $Mg_3[Si_4O_{10}](OH)_2$; 5 – base + 10% $Mg(OH)_2$; 6 – base + 5%TiO_2 + 5% $Mg_3[Si_4O_{10}](OH)_2$; 7 – base + 5% TiO_2 +5%$Al(OH)_3$. Designations: P – ammonium polyphosphate; A – aluminium hydroxide; T1 – titanium dioxide; T – talc; Mg – magnesium hydroxide

formation of aluminum orthophosphates of various structural types with the effect of reinforcing the porous foam coke layer (Figure 5, pos. b).

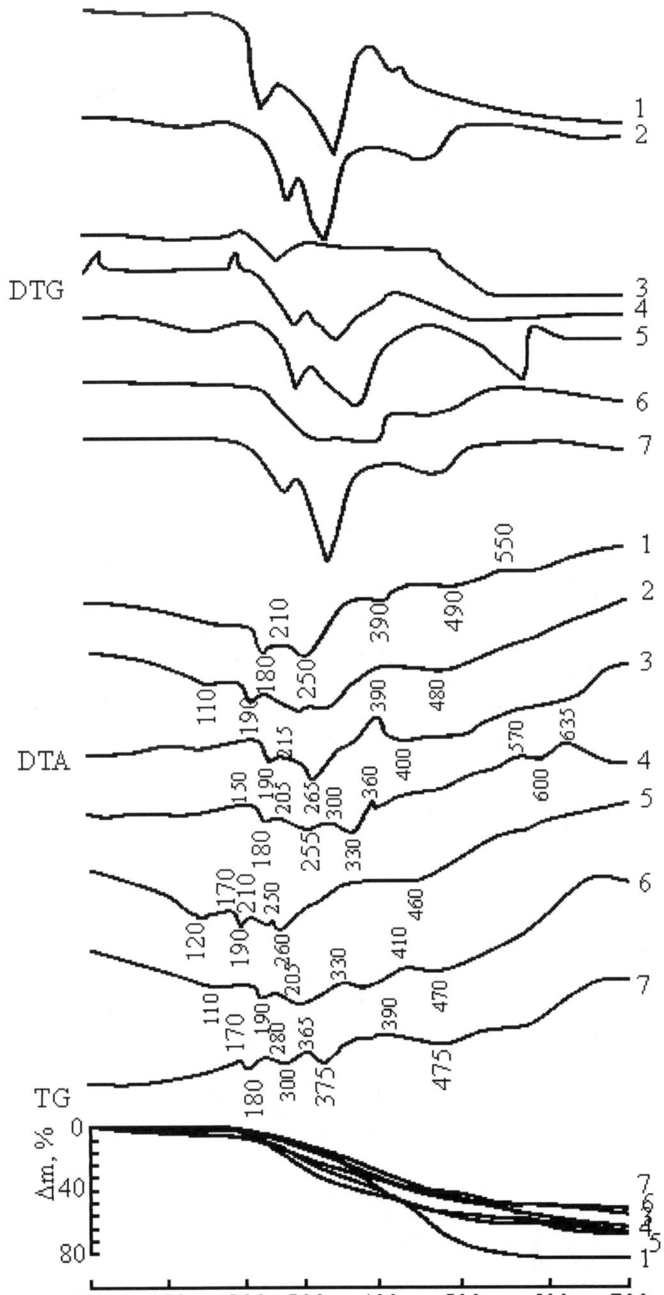

Figure 4. Thermograms of an organic-inorganic fire retardant composition: 1 – base composition; 2 – base + 10% $Al(OH)_3$; 3 – base + 10% TiO_2; 4 – base + 10% $Mg_3[Si_4O_{10}](OH)_2$; 5 – base + 10% $Mg(OH)_2$; 6 – base + 5%TiO_2 + 5% $Mg_3[Si_4O_{10}](OH)_2$; 7 base + 5% TiO_2 +5%$Al(OH)_3$

As a result of these processes, the coke foam structure is compacted and strengthened due to the formation of smaller pores compared to the previous type of structure (Figure 5, pos. A). In an organic-inorganic fire retardant composition containing up to 10% TiO_2 (Figure 3, curve 3), besides to the phases of ammonium polyphosphate, diffraction bursts are recorded, which are characteristic of TiO_2 (d =

0.2929; 0.249; 0.168; 0.154 nm). The XRD curve does not show diffraction reflections, which would be characteristic of titanium phosphates.

It is likely that titanium phosphates of the type $Ti(OH)_2(HPO_4)\ 2H_2O$ and $Ti(HPO_4)\ 2H_2O$ are formed as a result of the solid-phase reaction between TiO_2 and the product of partial dephosphorization of ammonium phosphate $(HPO_3)n$. With an increase in temperature from 300 to 600 °C on the DTA curve, an endothermic effect (–) 600 °C was noted, which is characteristic of the onset of dehydration of the above phases with the formation of anhydrous high-temperature titanium phosphates (Figure 4, curve 3), which significantly increase the thermal resistance of the coke foam layer (Figure 5, item c). The macrostructure of the coke foam layer consists of large globules in comparison with the macrostructure of the composition containing 10% $Al(OH)_3$.

When up to 10% talc ($Mg_3[Si_4O_{10}](OH)_2$) is added to the organic-mineral composition, diffraction bursts are recorded on the XRD curve (Figure 3, curve 4), which are characteristic of this mineral d = 0.266; 0.238; 0.175; 0.171; 0.159; 0.155 nm.

Dehydration of talc ($Mg_3[Si_4O_{10}](OH)_2$) in the direction of enstatite formation takes place at temperatures of 930-950 °C. Enstatite reacts with partially dephosphised (HPO_3) n to form magnesium silicophosphate. This phase is sufficiently temperature resistant and significantly increases the temperature resistance of the coke foam layer. The macrostructure of the coke foam is somewhat similar to the macrostructure of the base composition, but differs from it in its continuity (Figure 5, pos. D).

The introduction of up to 10% magnesium hydroxide into the base composition does not fundamentally change the formation of the macrostructure of the coke foam layer (Figure 5, pos. E). Dehydration $Mg(OH)_2$ d = 0.47; 0.179; 0.149 nm (Figure 1, curve 5) occurs in the temperature range 400-550 °C (Figure 4, curve 5) with the formation of MgO and its subsequent interaction with $(HPO_3)n$ in the direction of magnesium phosphate synthesis, as a high-temperature phase, which strengthens the structure coke layer and increases its resistance to high temperatures. Macrostructure of a scaly-type coke foam layer (Figure 5, pos. E).

The combined introduction of inorganic additives of titanium dioxide and talc in the amount of 5% does not lead to a significant change in the phase composition of the coke foam. The XRD curve (Figure 3, curve 6) shows diffraction bursts of ammonium phosphate, titanium dioxide (d = 0.293; 0.249; 0.168; 0.154 nm) and talc (d = 0.688; 0.33; 0.238; 0.190; 0.176; 0.142 nm). The macrostructure of the coke foam layer becomes denser due to its reinforcement with particles of titanium-magnesium phosphates (Figure 5, pos. F).

In our opinion, the most expedient is the introduction into the base, organic-mineral composition, a mixture of titanium dioxide with aluminum hydroxide in an amount of 5% each. On the XRD curve (Figure 3, curve 7) the diffraction bursts that are characteristic of ammonium polyphosphate and aluminum hydroxide (d = 0.309; 0.248; 0.229; 0.204; 0.169; 0.162; 0.145 nm) differ quite significantly. Diffraction bursts for titanium dioxide (d = 0.292; 0.262; 0.249; 0.2017; 0.168; 0.154 nm) are characterized by a lower intensity. With an increase in temperature in the composition of the products, the formation of mixed titanium-aluminum phosphates is possible, which differ from titanium-magnesium phosphates in increased heat resistance. The macrostructure of the coke foam layer is similar to the composition, containing up to 10% aluminum hydroxide (Figure 5, pos. B), but differs from it in a greater degree of ordering (Figure 5, pos. G).

Taking into account the data of physicochemical studies, it is possible to propose a model of the mechanism of structure formation of an expanded coke foam layer and the main directions of control of this process based on fuzzy mathematics [44,45].

IOP Conf. Series: Materials Science and Engineering 1162 (2021) 012003 doi:10.1088/1757-899X/1162/1/012003

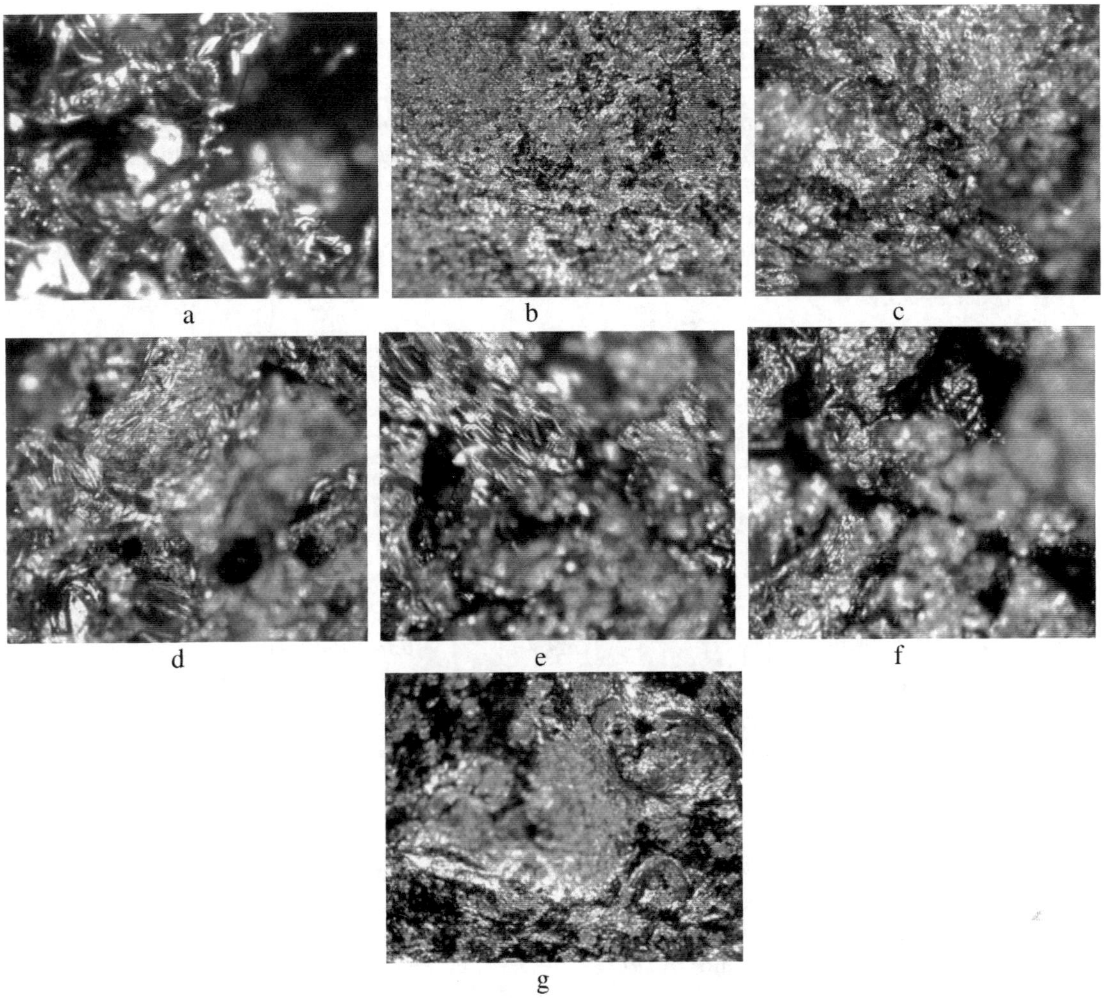

Figure 5. Digital photos (x500) of the coke foam layer after swelling at a temperature of 500 °C: a – base composition; b – base + 10% $Al(OH)_3$; c – base + 10% TiO_2; d – base + 10% $Mg_3[Si_4O_{10}](OH)_2$; e – base + 10% $Mg(OH)_2$; f – base + 5%TiO_2 + 5% $Mg_3[Si_4O_{10}]$ $(OH)_2$; g – base + 5% TiO_2 +5%$Al(OH)_3$.

Table 2. Compounds of organic-mineral fire retardant compositions

№	Composition	Amount and type of additive, %	Swelling coefficient value
1	Base	–	26.8
2	Base +	10% $Al(OH)_3$	36.7
3	Base +	10% TiO_2	17.2
4	Base +	10% $Mg(OH)_2$	32.1
5	Base +	10% $Mg_3[Si_4O_{10}](OH)_2$	30.1
6	Base +	5%TiO_2+5% $Al(OH)_3$	47.0
7	Base +	5%TiO_2+5% $Mg_3[Si_4O_{10}](OH)_2$	42.0

A fuzzy model of the structure formation process can be presented in the following form:

$$\mu^k(lt) = \cap_{j=1}^{12} \cup_{i+1}^{11} \left(x_i \to x_y\right), \quad (1)$$

where $\mu^k(lt)$ – the measure of belonging of the linguistic variable *lt*, which characterizes the development of the process; *lt* matters: the speed of the process, the type of process, the type of materials and equipment; *k* – type of object of protection (wooden constructions of industrial, residential premises; x_i = 1 ... 5 – input linguistic variables (table 3); x_y = 6 ... 12 – input linguistic variables (table 4).

The basis for fuzzy inference is a knowledge base containing a set of fuzzy production rules (FPRs) that determine the strategy for solving the structure formation process and the ability to control it. A typical rule base production rule consists of a premise (antecedent): fuzzy statements in the form "if" and a conclusion (consequent) in the form "then ...".

Table 3. An example of formalization of direct parameters of the structure formation process of an organic-inorganic fire retardant composition.

Process types	Terms for linguistic evaluation
x_1 – structure organization processes	diffusion (df); coagulation (cg); solid phase (sp)
x_2 – process development	constant intensity (ci); variable decelerating intensity (vdi); variable accelerating intensity (vai)
x_3 – types of processes	material (m); technological (t)
x_4 – life cycle of processes	inception (in); acceleration (ac); deceleration (dc); age (ag); attenuation (at)
x_5 – interaction of processes	component composition set (ccs); reactionary components set (rcs); structure formation period (sfp)

An antecedent can contain several premises, which are combined depending on the strategy using logical connectives "and" or "or". Each of the rules of fuzzy rules can have some weight *Fi* = *[0,1]*, which determines the significance of the rule or confidence in the degree of truth of the conclusion obtained by a separate fuzzy rule. In general, the base containing m rules is as follows: R1: if *x1* is *A11* ... and (or) ... *xn* is *A1n*, then *y* is *R1*; *Ru*: if *x1* is *Ai1* ... and (or) ... *xn* is *Ain*, then *y* is *Ru*; *Rm*: if *x1* is *Ai1* ... and (or) ... *xn* is *Amn*, then *y* is *Rm*, where: *xk* – input variables, *k=1, ..., n*; *y* – output variable; *Aik* – given fuzzy sets with membership functions (*i = 1, ..., m; k = 1, ..., n*).

Table 4. An example of formalization of indirect parameters of the structure formation process.

Diagnostic parameters of incoming data	Terms for linguistic evaluation
x_6 – influence of temperature and humidity on the reaction rate	insignificant (is), significant (s)
x_7 – component view of materials	common (c), innovative (in)
x_8 – type of equipment	standard (sd), non-standard (nsd)
x_9 – cost of materials	low (l); medium (m), high (h)
x_{10} – hardening time	slow (sl), normal (nl), fast (ft)
x_{11} – influence of fire factors	short-term (st), long (lg)
x_{12} – influence of rheological factors	viscous (v), normally viscous (nv), fluid (fd)

When using fuzzy clustering algorithms (formula 1), if the weight coefficients or the degree of influence of each significant factor of external conditions are not determined, the result of the algorithm depends on the order of the diagnostic parameters that form the input data vector. That is why, in order to reduce the level of risk of making a wrong decision, it is proposed to take into account the degree of influence of various environmental factors when forming the vector of input data using their sorting.

In case of a conflict of rules, the fuzzy system will recommend making decisions according to the criterion with a lower ordinal number (in the absence of other rules). On the basis of fuzzy statements, the truth of which is established as a result of fuzzification, the degree of truth of fuzzy statements is assessed, and is the conclusion of the corresponding FPRs.

Next, a procedure (aggregation) is performed to determine the degree of truth of the left parts (cut-off levels – ai) for each of the rules of the fuzzy inference system using the FUZZY EXPERT software [46, 47].

6. Conclusions

The peculiarities of structure formation of intumescent type organic-mineral paints for fire protection of wooden structures have been studied. On the basis of research, a scheme of the coke formation process has been developed. It has been shown that the introduction of mineral fillers in the composition of organic-mineral paints in an amount of 10% promotes an increase in the swelling coefficient from 30 to 36.7, which is 1.5 ... 1.84 times higher than the values of the swelling coefficient of the base organic-mineral paint. The introduction of up to 10% TiO_2 contributes to a decrease in the swelling coefficient by a factor of 1.16 in comparison with the base one, but significantly increases the thermal stability of the formed coke foam. The introduction of titanium oxide and aluminum hydroxide in an amount of 5% promotes the ordering of the structure of the foam layer, an increase in its thermal stability and the achievement of the maximum value of the swelling coefficient – up to 47, which is 1.11 times and 1.7 times higher than the values of the swelling coefficient with the introduction of aluminum hydroxide and basic compositions. The introduction of a mixture of titanium oxide and talc in an amount of 3.3% increases the values of the swelling coefficient by 1.12 times in comparison with the introduction of titanium oxide and talc in an amount of 5%, and 1.08 times more than the values of the swelling coefficient with the introduction of aluminum hydroxide and talc in the amount of 5%. It was found that organic-mineral paint at high temperatures is capable of significant weight loss, namely at 700 °C due to the formation of refractory compounds of the type TiP_2O_7, $Mg_3(PO_4)_2$, $AlPO_4$, which prevent the burnout of the formed coke foam. Taking into account the data of physicochemical studies, on the basis of fuzzy mathematics, a model of the mechanism of structure and coke formation and control of this process are proposed.

Determining the fire retardant ability and effectiveness for passive fire retardant coatings will increase the competitiveness and reduce the cost of such materials. In order to ensure the necessary fire resistance of the structure, determined in accordance with the current regulatory documents, wooden constructions are subjected to fire retardant treatment – an important and expensive stage of construction. For the owner of the building, effective fire protection makes it possible to ensure the safe evacuation of people from the building, preserve material values and create an opportunity for the safe work of rescue units [48].

References

[1] Kovalov A, Otrosh Y, Vedula S, Danilin O, Kovalevska T 2019 Parameters of fire-retardant coatings of steel constructions under the influence of climatic factors *Scientific Bulletin of the National Mining University* **3** pp 46–53

[2] Vasilchenko A, Otrosh Yu, Adamenko N, Doronin E, Kovalov A 2018 Feature of fire resistance calculation of steel structures with intumescent coating *MATEC Web of Conferences* **230**, 02036

[3] Kovalov A, Otrosh Y, Kovalevska T, Safronov S 2019 Methodology for assessment of the fire-resistant quality of reinforced-concrete floors protected by fire-retardant coatings *In Materials Science and Engineering IOP Publishing* **708** (1) p 012058

[4] Krivenko P, Pushkarjeva E, Sukhanevich M, Guziy S 2009 Fireproof coatings on the basis of alkaline aluminum silicate systems Developments in Strategic Materials: *Ceramic Engineering and Science Proceedings* **29** **10** pp 129–142

[5] Tsapko Yu, Guzii S, Remenets M, Kravchenko A, Tsapko A 2016 Evaluation of effectiveness of wood fire protection upon exposure to flame of magnesium *Eastern-European Journal of Enterprise Technologies* **4(10-82)** pp 31-36

[6] Guzii S, Tsapko Yu, Kravchenko A, Remenets M 2016 Fire Protection of wooden storage containers for explosive and pyrotechnic products *Eureka: Physics and Engineering Sciences*, **2**, pp 34–42

[7] Krivenko P, Guzii S, Kravchenko A 2013 Protection of Timber from Combustion and Burning Using Alkaline Aluminosilicate-Based Coatings *Advanced Materials Research* **688** pp 3-9

[8] Guzii S, Bazhelka I, Svitlychna N, Lashchivskiy V 2020 Protection of Wood from Burning with Paints on Alkaline Aluminosilicates-Based *Materials Science Forum* **1006** pp 19-24

[9] Guzii S, Krivenko P, Kravchenko A 2012 Protection of wood from flammability by coatings based on alkaline hydroaluminosilicates. Scientific and technical collection *Building materials, products and sanitary ware* **45** pp 38-43

[10] Guzii S, Tsapko Yu 2013 The influence of modification of wood on the fire resistance of wooden structures *Materials 13th International Scientific Conference VSU'*, Sofia, Bulgaria, 6-7 June pp IV-29-IV-35

[11] Tsapko Yu, Kravchenko A, Tsapko O 2017 Analysis of methods and development of a method for determining the effectiveness of fire protection of wooden structures. *Bulletin of the Odessa State Academy of Civil Engineering and Architecture* **66** pp 118-122

[12] Chou CS, Lin SH, Wang CI, Liu KH 2010 A hybrid intumescent fire retardant coating from cake- and eggshell-type IFRC *Powder Technology* **198(1)** pp 149-156

[13] Bourbigot S, Gardelle B, Duquesne S 2014 Intumescent silicone-based coatings for the fire protection of carbon fiber reinforced composites *Fire safety science-proceedings of the eleventh international symposium* pp 781-793

[14] da Silveira M R, Peres R S, Moritz V F, Ferreira C A 2019 Intumescent Coatings Based on Tannins for Fire *Protection Materials Research* **22(2)**: e20180433

[15] Oliveira R B RS, Moreno Junior A L, Vieira L C M 2017 Intumescent paint as fire protection coating *Ibracon Structures and Materials Journal* **10 1** pp 220-243

[16] Andreza P Cardoso, Stéphanie C de Sá, Carlos H M. Beraldo, Gelsa E N Hidalgo & Carlos A 2020 Ferreira Intumescent coatings using epoxy, alkyd, acrylic, silicone, and silicone–epoxy hybrid resins for steel fire protection *Journal of Coatings Technology and Research* **17** pp 1471–1488

[17] Gao M, Yang S 2010 A novel intumescent flame-retardant epoxy resins system *Journal of Applied Polymer Science* 115(4):2346- 2351

[18] Sakhno T, Barashkov N, Irgibaeva I, Pustovit S, Sakhno Yu E 2016 Polymer Coatings for Protection of Wood and Wood-Based Materials *Advances in Chemical Engineering and Science* **6** pp 93-110

[19] Danchenko Y, Andronov V, Barabash E, Obigenko T, Rybka E, Meleshchenko R, Romin A 2017 Research of the intramolecular interactions and structure in epoxyamine composites with dispersed oxides *Eastern-European Journal of Enterprise Technologies* **6 12–90** pp 4–12

[20] Danchenko Y, Andronov V, Teslenko M, Permiakov V, Rybka E, Meleshchenko R, Kosse A 2018 Study of the free surface energy of epoxy composites using an automated measurement system *Eastern-European Journal of Enterprise Technologies* **1 12–91** pp 9–17.

[21] Tsapko Yu, Kravchenko A, Tsapko O 2017 The effect of adding aluminum hydrate on the effectiveness of fire protection of wood with organo-mineral coating *Scientific and technical journal "New technologies in construction"* 33 pp 46-49

[22] Tsapko Yu, Kravchenko A, Guziy S, Tsapko O 2017 Structure formation of the expanded layer of foam coke and effective measures for its management. Scientific and technical collection *Building materials, products and sanitary ware* **58** pp 171-179

[23] Givlyud M 2011 Temperature-resistant silicate protective coatings for metals and alloys based on filled polymethylphenylsiloxane *Collection of scientific works* of Lviv State University BJD 18 pp 40-45

[24] Eremina T Yu, Gravit M V, Dmitrieva T Yu 2012 Features and principles of constructing formulations of fire retardant expanding compositions based on epoxy resins *Fire and explosion safety J* **21 7** pp 52-56
[25] Jimenez M, Bellayer S, Naik A, Bachelet P, Duquesne S, Bourbigot S 2016 Topcoats versus Durability of an Intumescent Coating *Industrial and Engineering Chemistry Research* **55(36)** pp 9625-9632
[26] Zhong Y, Wu W, Wu R, Luo Q, Wang Z 2014 The flame retarding mechanism of the novolac as char agent with the fire retardant containing phosphorous-nitrogen in thermoplastic poly(ester ether) elastomer system *Polymer Degradation and Stability* **105** pp 166-177
[27] Ullah S, Ahmad F 2014 Effects of zirconium silicate reinforcement on expandable graphite based intumescent fire-retardant coating *Polymer Degradation and Stability* **103** pp 49-62
[28] Bouvet N, Linteris G, Babushok V, Takahashi F, Katta V, Krämer R 2016 Experimental and numerical investigation of the gas-phase effectiveness of phosphorus compounds *Fire and Materials* **40(5)** pp 683-696
[29] Mashlyakovsky L N, Lykov A D, Repkin Yu V 1989 Low flammability organic coatings Leningrad *Chemistry*, 184
[30] Mashlyakovsky L, Lykov A, Aleskerov M 1992 *J. Paints and varnishes* **6** pp 45–51
[31] Nenakhov S, Pimenova V 2011 Dynamics of foaming fire-retardant coatings based on organic-inorganic compositions Fire and explosion safety **20 8** pp 17-24
[32] Nenakhov S, Pimenova V 2010 The influence of the concentration of the gassing agent on the patterns of development of coke foam in flame retardants *Fire and explosion safety* **19 3** pp 14-26
[33] Nenakhov S, Pimenova V 2010 Physicochemistry of foaming fire-retardant coatings based on ammonium polyphosphate (literature review) *Fire and explosion safety* **19 8** pp 11-58
[34] Bazhenov S et al 1999 RF Patent № 2180741, IPC G01N3/08 Installation for testing fire-retardant swellable coatings on metal *Application 04.08.1999, Publ. 20.03.2002*; *Bull.* **12** p 10
[35] Tsapko Yu 2012 Research of separate parameters of destruction of wood at operation. *Building constructions of buildings and constructions: designing, manufacturing, reconstruction and service* Bulletin of DNABA **5 (97)** pp 56-61
[36] GOST 18992-80 Dispersion polyvinyl acetate homopolymer coarse. Specifications 2001 Moscow *Publishing House of Standards* p 17
[37] Ammonium polyphosphate. *Technical conditions*. TU 2148-029-13267785-04 2004 p 18
[38] GOST 7579-76 Melamine. Specifications 1998 Moscow *Publishing House of Standards* p 6
[39] GOST 9286-89 Technical pentaerythritol. Specifications. Moscow 2001 *Publishing House of Standards* p 12
[40] Stoye D, Freytag W 2007 Paints, coatings and solvents SPb *Profession* p 528
[41] Gorshkov V, Timashev V, Savelyev V 1984 Methods of physicochemical analysis of binders. Textbook Moscow *Higher school* p 335
[42] Wladyka-Przybylak M, Kozlowski R 1999 Thermal characteristics of different intumescent coatings *Fire and Materials* 23(1) pp 33- 43
[43] DSTU-N-P B B.1.1-29: 2010 Fire protection. Fire-retardant treatment of building structures. General requirements and control methods. *Kyiv State Enterprise "State Research Institute of Building Structures" (NDIBK)* p 9
[44] Teslenko A, Chernukha A, Bezuglov O, Bogatov O, Kunitsa E, Kalyna V, Katunin A, Kobzin V, Minka S 2019 Construction of an algorithm for building regions of questionable decisions for devices containing gases in a linear multidimensional space of hazardous factors *Eastern-European Journal of Enterprise Technologies* **5 10–101** pp 42–49
[45] Otrosh Y, Rybka Y, Danilin O, Zhuravskyi M 2019 Assessment of the technical state and the possibility of its control for the further safe operation of building structures of mining facilities *EDP Sciences In E3S Web of Conferences* **123** p 01012
[46] Otrosh Y, Kovalov A, Semkiv O, Rudeshko I, Diven V 2018 Methodology remaining lifetime

determination of the building structures *MATEC Web of Conferences* **230** p 02023
[47] Shnal T, Pozdieiev S, Sidnei S, Shvydenko A 2021 Determination of the Charring Rate of Timber to Estimate the Fire Resistance of Structures at Real Temperature Modes of Fires *Lecture Notes in Civil Engineering* 100 LNCE pp 409–418
[48] Tiutiunyk V, Ivanets H, Tolkunov I, Stetsyuk E 2018 System approach for readiness assessment units of civil defense to actions at emergency situations *Scientific Bulletin of the National Mining University* **1** pp 99–105

Modeling of the stress-strain state of reinforced concrete beams under prolonged load action

Z Holovata[1,2], S Neutov[1], M Surianinov[1]

[1] Odessa State Academy of Civil Engineering and Architecture, 4, Didrikhsona st., Odessa 65029, Ukraine
[2] Address all correspondence to this author.

E-mail: zlataholovataya@gmail.com

Abstract. The paper considers the use of a layered deformation model of the stress-strain state of reinforced concrete beams experiencing transverse bending taking into account the long-term action of the load. The results of experimental studies of twelve beams made of concrete of three classes for short-term and long-term acting load are given. It has been established that long-term strength of inclined cross-sections follows the general laws of deformation and fracture of reinforced concrete. Comparison of calculation results of the proposed deformation model with experimental data has shown their satisfactory convergence. A tendency of some increase (within 10...15%) in the long-term strength of the reinforced concrete elements under study in the inclined sections has been established experimentally.

1. Introduction

There are many works devoted to the study of reinforced concrete beams for strength, stiffness and cracking resistance under prolonged exposure to load. Much fewer publications are connected with the consideration of this problem in relation to fiber reinforced concrete beams.

Long-term exposure to loading that causes microcracks in the material during initial loading shows that the failure of the specimen proceeds slowly over a long period of time. Experiments with concrete (and to an even greater extent with fiber concrete) show that the long-term strength of concrete lies above the limit that determines the strength under long-term loading. There is still no consensus on the factors influencing the long-term strength of concrete and fiber concrete beams when they are bent. The problem requires further research and continues to be relevant. Nowadays in Ukraine the design of concrete and reinforced concrete structures is carried out according to the limit states of the first and the second groups taking into account as a rule the buildings and structures responsibility classes and responsibility categories of structural elements set by corresponding normative documents, variability of properties, loads and influences, geometry characteristics, operation term and working conditions of structures [1]. Calculation of fiber concrete structures when reinforced with steel fiber is regulated by the normative document DSTU-N B B.2.6-218:2016 [2].

2. Literature review

The existing methods of calculation of reinforced concrete elements under the action of bending moments, longitudinal and transverse forces can be divided into two main groups: methods based on consideration of the stress-strain state of the section [3-6] and methods not using strain relations. The first group usually includes the methods of calculation of elements based on normal sections, contained in the design standards of most developed countries, including the European standards [7-9]. These methods are based on calculation models based on unified principles, the main of which are the hypothesis of plane sections and material deformation diagrams. This approach makes it possible to calculate strength, crack formation and crack opening, as well as deformations from a unified position. Characteristic representatives of the second group are methods of calculation, which are in good agreement with experiments, but do not have a common basis in the calculations for various types of limit states. These methods are full of numerous empirical relations.

The study of bendable reinforced concrete elements under long-term loads was carried out by many well-known scientists. However, in our country in recent years, such studies were carried out little. The work [10] is devoted to the study of cracking in reinforced concrete beams under long-term loads on the basis of the energy approach. Development of calculation methods for steel reinforced concrete beams as elements of slabs under different types and modes of loading is devoted to the article [11].

In foreign publications, the emphasis has shifted to the study of beams with dispersed reinforcement. Thus, the paper [12] presents the results of studying the long-term behavior of composite reinforced concrete beams under long-term loads. It is noted that the deflections in the middle of the span after 3 years under load increased by 2,5 times compared to the initial deflections. Work [13] is aimed at studying, both analytically and experimentally, the deflections of beams reinforced with fiberglass that have been subjected to prolonged loads for 2 years. Experiments with beams made of high-strength fiber reinforced concrete (SFRC) subjected to long-term bending loads are described in [14].

3. The purpose of this work is to use a layered deformation model of the stress-strain state of reinforced concrete beams experiencing transverse bending, taking into account the long-term action of the load.

4. Materials and Methods
Mathematical methods of experiment planning with obtaining and analysis of appropriate experimental and statistical models have been used. Determination of physical, physical-mechanical and construction-technical properties of concrete and fiber concrete has been carried out in accordance with the current regulatory documents and generally accepted methods.

5. Results and Discussion
We assume the deformation model of the beam in discrete form: we divide the beam into i sections (multiples of the step of the transverse reinforcement) along the length and j layers (1 cm each) along the height of the section. The stress-strain state of the formed sections is assumed to be the same.

The reinforcement of the tensile and compressed zones of the beam is "tied" to the faces of the section at the level of their centers of gravity with the values of the protective layers a and a' (figure 1).

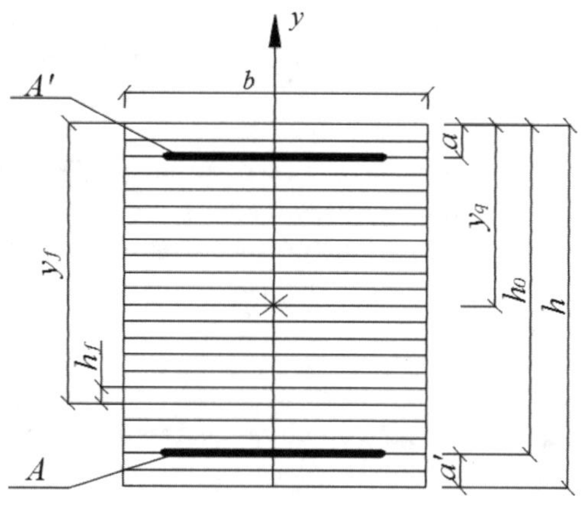

IOP Conf. Series: Materials Science and Engineering 1162 (2021) 012004 doi:10.1088/1757-899X/1162/1/012004

Figure 1. Design cross-section.

The modeling of the stress-strain state of the beam, experiencing transverse bending under prolonged load action, is based on the following: the relations between the stresses and strains of concrete and reinforcement are represented by real diagrams $\sigma - \varepsilon$; distribution of deformations along the height of sections occurs in accordance with the hypothesis of plane sections; the hypothesis of joint deformations of concrete and reinforcement is accepted; the loading of the beam is carried out by two symmetrically arranged forces F, applied at a distance c from the supports (figure 2); all characteristics of the stress-strain state of the beam are determined by the method of successive approximations.

To describe diagram $\sigma_b - \varepsilon_b$ (figure 3), we use relation:

$$\frac{\sigma_b(t)}{R_b} = \frac{k \cdot \eta - \eta^2}{1 + (k-2) \cdot \eta^3}, \tag{1}$$

where $\eta = \varepsilon_b(t)/\varepsilon_{bu}$ – value, which characterizes the intensity of the increase in deformation in concrete $\varepsilon_b(t)$; ε_{bu} – the deformation value that corresponds to the greatest value of prism strength R_b; k – a coefficient that takes into account the influence of the inelastic component of the deformation.

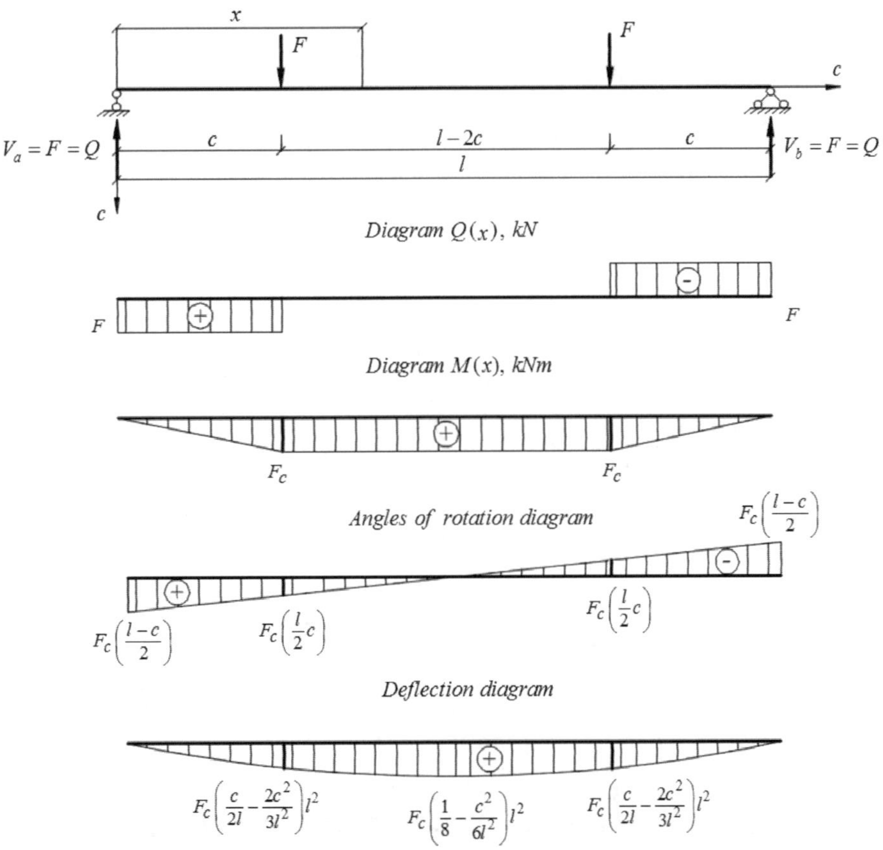

Figure 2. Diagrams of internal forces, rotation angles and deflections of the beam from the action of the transverse load without taking into account its own weight

The prism strength of concrete at any point in time t from the day of concreting is determined by the known formula:

$$R_b = R_{b28}\frac{\lg t}{\lg 28}, \tag{2}$$

where R_{b28} – prism strength of concrete at 28 days of age.

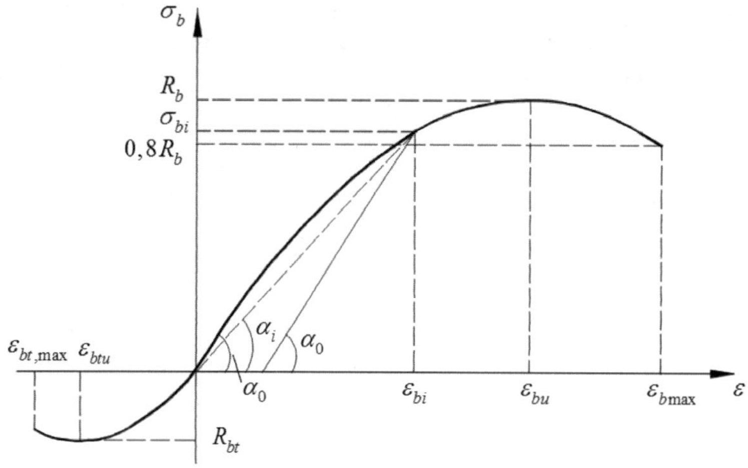

Figure 3. Diagram $\sigma - \varepsilon$ of concrete

To determine the value ε_{bu} we accept the formula, which according to [15, 16] agrees well (variation coefficient – $\nu \le 5\%$) with experimental data:

$$\varepsilon_{bu}\left[3,6R_b - \left(3\frac{R_b}{36}\right)^5 + 51\right]\cdot 10^{-5}. \tag{3}$$

In order to bring the ascending branch of the curve (1) closer to the experimental data, an additional condition was introduced when determining the coefficient k . Such a condition is the point of the diagram, which corresponds to the initial deformation modulus of concrete E_{b0} . In the current norms in Ukraine [1], the initial modulus E_{b0} is determined at stresses $\sigma_b = 0,3R_b$. For this stress level, the deformation of the concrete is

$$\varepsilon_b = \frac{\sigma_b}{E_{b0}} = \frac{0,3R_b}{E_{b0}},$$

then

$$\eta_0 = \frac{0,3R_b}{\varepsilon_{bu}E_{bo}}. \tag{4}$$

After substitution into (1) $\sigma_b/R_b = 0,3$ and expression (4) we have:

$$0,3 = \frac{k\cdot\eta_0 - \eta_0^2}{1+(k-2)\eta_0}. \tag{5}$$

After transforming (5), we get the expression for the coefficient of elasticity of concrete k on the ascending branch of the diagram:

$$k = \frac{\eta_0^2 + 0,6\eta_0 + 0,3}{0,7\eta_0}. \tag{6}$$

To describe the descending branch of the concrete compression diagram, the point corresponding to the final value of the concrete strain $\varepsilon_{b\max}$ is introduced:

$$\varepsilon_{b\max} = \frac{R_b \cdot 10^{-2}}{10 + 2,75R_b}. \quad (7)$$

On the assumption that $\varepsilon_{b\max}$ is observed at stresses $\sigma_b = 0,8R_b$, coefficient of elasticity k_1 on the downward branch of the diagram $\sigma - \varepsilon$ of the concrete takes the form of

$$k_1 = \frac{\eta_{01}^2 + 1,6\eta_{01} + 0,8}{0,2\eta_{01}}. \quad (8)$$

To describe diagram $\sigma - \varepsilon$ for tensile concrete we use the same dependence (1) as for compression. Values ε_{btu}, $\varepsilon_{bt\max}$ we obtain by multiplying ε_{bu}, $\varepsilon_{b\max}$ on the ratio R_{bt}/R_b. In the final form, the tensile stresses in the concrete are determined using the dependence

$$\frac{\sigma_{bt}(t)}{R_{bt}} = \frac{k \cdot \eta_t + \eta_t^2}{1 + (k-2)\eta_t}, \quad (9)$$

where $\eta_t = \varepsilon_{bt}(t)/\varepsilon_{btu}$ – value that characterizes the intensity of tensile deformation ε_{bt} in concrete; ε_{btu} – deformation value of concrete at maximum tensile stresses.

The tensile strength of concrete is determined by the formula

$$R_{bt} = R_{bt28} \frac{\lg t}{\lg 28}. \quad (10)$$

The secular modulus of deformation of concrete at any age and level of loading at some point i of the diagram is defined by the dependence

$$E_{bi}(t) = \frac{\sigma_{bi}(t)}{\varepsilon_{bi}(t)} = tg\alpha_i. \quad (11)$$

Creep deformations ε_{cr} can be taken into account by using the known dependences proposed in [5, 6] for concretes without damage and in [15, 16] for concretes damaged by corrosive processes. The determination of creep deformations [15] is based on the law of change in the value of concrete creep characteristic suggested by A. B. Golyshev [5], which can be represented in the following form:

$$\varphi(t) = \varphi_k (1 - e^{-\lambda(t)}), \quad (12)$$

where $\lambda(t) = 1,5\sqrt{t/365}$; φ_k – the final value of the creep characteristic, which can be determined taking into account the corrosion damage of concrete by the formula:

$$\varphi_k = \xi_b^c \cdot \xi_{cr1} \cdot \xi_{cr2} \cdot \xi_{cr3} \cdot C_m \cdot E_b. \quad (13)$$

The normative value of the creep measure can be determined by the formula

$$C_m = 12,5 \cdot 10^{-6} \frac{B}{R_b}. \quad (14)$$

Conclusively, given that $\varphi(t) = \varepsilon_{cr}(t)/\varepsilon_b(t)$, after some transformations, the formula for determining the creep deformation of concrete with regard to corrosion damage will look like

$$\varepsilon_{cr}(t) = \xi_b^c \cdot \xi_{cr1} \cdot \xi_{cr2} \cdot \xi_{cr3} \cdot 12,5 \cdot 10^{-6} \frac{B \cdot \sigma_b(t)}{R_b}(1 - e^{\lambda(t)}), \quad (15)$$

where ξ_b^c – coefficient, which takes into account the presence of corrosion damage and which is recommended to take equal to 1,05 ... 1,15, depending on the type of concrete; ξ_{cr1}, ξ_{cr2}, ξ_{cr3} – coefficients that take into account the humidity of the environment, the size of the structure, the time of their loading.

According to the recommendations of O. Y. Berg [17], they can be determined by the formulas:

$$\xi_{cr1} = \frac{1,54(135-W)}{100}; \tag{16}$$

$$\xi_{cr2} = 0,90\left(0,70\frac{1}{r}\right); \tag{17}$$

$$\xi_{cr3} = 0,36\frac{26,4}{(13+\tau)}, \tag{18}$$

where τ – loading time; r – exposed concrete specific surface; W – ambient humidity.

Swelling deformations of concrete $\varepsilon_{sh}(t)$ can be determined using the dependencies [5], as well as taking into account the effect of corrosion [15].

The value of shrinkage deformations of concrete according to [5] is determined by the formula

$$\varepsilon_{sh}(t) = \varepsilon_{shk}(1-e^{-\lambda(t)}), \tag{19}$$

where ε_{shk} – the final value of concrete shrinkage deformations, which, taking into account corrosion damage, can be determined from the condition:

$$\varepsilon_{shk} = \xi_b^c \cdot \xi_{sh1} \cdot \xi_{sh2} \cdot \xi_{shu}. \tag{20}$$

At the same time

$$\varepsilon_{shu} = 0,14 \cdot 10^{-6}\sqrt{B}; \tag{21}$$

$$\varepsilon_{sh}(t) = \varepsilon_{bc} \cdot \varepsilon_{sh1} \cdot \varepsilon_{sh2} \cdot 0,14 \cdot 10^{-6}\sqrt{B}(1-e^{-\lambda(t)}). \tag{22}$$

The final formula for determining the deformations takes the form

$$\varepsilon'_{sh}(t) = \gamma'_{sh}\varepsilon_{sh}(t), \tag{23}$$

where γ'_{sh} – transition coefficient from shrinkage deformation to swelling deformation; ε_{sh1}, ε_{sh2} – coefficients, which take into account the effect of humidity and size of structures and which [18] are recommended to determine from the conditions:

$$\varepsilon_{sh1}(t) = 1,52\left[1-\left(\frac{W}{100}\right)^3\right], \tag{24}$$

$$\varepsilon_{sh2}(t) = 0,035(31-r). \tag{25}$$

The total design deformations of concrete, taking into account long-term processes and corrosion, are:

$$\varepsilon_{bij}(t) = \varepsilon_{bijo} + \varepsilon_b(\sigma_b, t) + \varepsilon_{cr}(t) - \varepsilon_{sh}(t)_q, \tag{26}$$

where ε_{bijo} – deformations of concrete under short-term loading; $\varepsilon_b(\sigma_b, t)$ – deformations of concrete due to stress growth in the specimens due to a decrease in the cross-section due to corrosion stresses.

In [19] it was proposed to take into account the transformation of the parameters of the deformation diagram of concrete depending on the time and level of long-term loading, based on the values of these parameters during short-term loading using the following dependencies:

$$\xi_{b\max}(t)=\xi_{br}[1+\eta_{\varepsilon}(0,1800+0,2734\, ln(t))]; \quad (27)$$

$$\xi_{bu}(t) = \xi_{bu}[1+ \eta_{\varepsilon}(0.206+0.239 \ln(t))]; \quad (28)$$

$$\beta(t)=\beta+\left(\frac{\xi_{bu}(t)-\xi_{bu}}{\xi_{bu}}\right)\left(\frac{0,932-\beta}{2}\right); \quad (29)$$

$$\sigma_{b\max}(t)=R_b[0,9488-0,0166\ln(t)]\, k\,(\eta_{\varepsilon},t); \quad (30)$$

$$k\,(\eta_{\varepsilon},t)=1-0,152\ln(t)[1-1,4\,\eta_{\varepsilon}+0,491\,\eta_{\varepsilon}^{2}]. \quad (31)$$

where η – long-term stress level; $R_b(\tau-\tau_1)$ – prism strength of concrete after prolonged stress action; $R_b(\tau)$ – is the prism strength of the unloaded concrete specimen at the considered moment of time; τ – stress action duration.

The relationship between the stresses and deformations of concrete under prolonged load action, as recommended in [19], can be taken as a diagram described by a polynomial

$$\sigma_b(t)=\sigma_{b\max}(t)\sum_{i=1}^{5} a_i(t)\left(\frac{\xi_b(t)}{\xi_{b\max}(t)}\right)^{k}, \quad (32)$$

where $\sigma_b(t)$ – stress in the concrete at an arbitrary moment of time; $\xi_b(t)$ – total deformations of concrete at an arbitrary moment of time; $a_i(t)$ – coefficients of the polynomial describing the deformation diagram of concrete under long-term load action.

The relationship between stresses and strains is represented in the form of two linear diagrams (figure. 4).

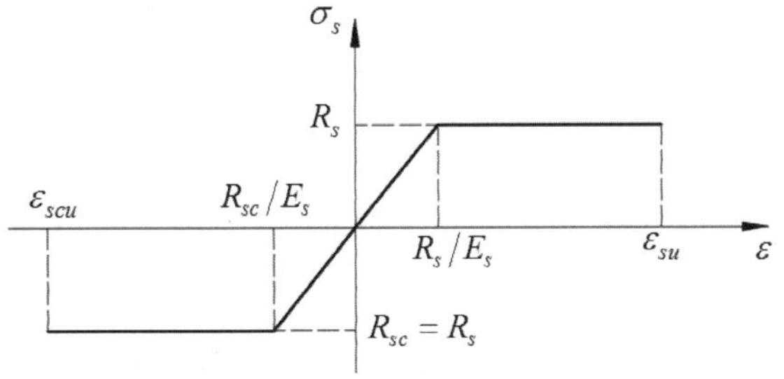

Figure 4. Diagram $\sigma_s-\varepsilon_s$ for reinforcing steel

The bending moment and shear force along the length of the element are variables:

$$\begin{aligned} &-at\ 0<x\le c && M(x)=Fx;\ Q(x)=F; \\ &-at\ c<x<l-c && M(x)=Fc;\ Q(x)=0; \\ &-at\ l-c<x\le l && M(x)=F(l-x);\ Q(x)=-F. \end{aligned} \quad (33)$$

Experimental studies have been carried out for the long-term action load of three series of specimens of 4 beams in each with the shear span $c=35cm$, made of concrete class C12/15, C20/25, C28/35, respectively. The destruction of all the beams, both under short-term and long-term action of the load, occurred along the inclined crack. In each series, one beam was subjected to short-term loading until failure, and the remaining 3 beams were subjected to long-term loading, respectively, with load levels of 0,85; 0,9 and 0,95 of the destructive load. The experiments showed that the beams with a load level of 0,95 failed, depending on the concrete class, in the interval from 1 to 65 hours. The higher the concrete class, the longer the interval to failure. Beams loaded to the 0,9 level of breaking load stood under load for 5 to 70 days, after which failure occurred. When tested at this load level, a steady increase in deflections and crack opening widths was observed. By the time of destruction, the width of inclined cracks opening under prolonged action of load exceeded the similar indicators under short-term load in 1,6–2,1 times, and the deflections - in 1,7–2 times.

In beams of all three classes of concrete at a load of 0.85Fu, stabilization of deflections and crack opening width were observed. After an eight-month observation period, these beams were brought to failure. As a result of studies, it was found that the failure load for beams loaded with $0,85F_u$ long-term load increased by 13-15% compared to short-term loading.

6. Conclusions

Thus, it has been established that the long-term strength of inclined sections obeys the general laws of deformation and fracture of reinforced concrete.

Comparison of the results of calculations using the proposed deformation model with experimental data showed their satisfactory convergence.

The tendency of some growth (in the range of 10...15%) of the long-term strength of the studied reinforced concrete elements in inclined sections has been established experimentally.

References

[1] DBN V.2.6-98: 2009 2001 *Concrete and reinforced concrete structures. The main provisions* Ministry of Regional Development of Ukraine p 73

[2] DSTU-N B V.2.6-218:2016 2017 *Regulations on the design and manufacture of structures made of dispersion-reinforced concrete* Kyiv DP UkrNDNTs p 35

[3] Zalesov A, Chystiakov E, Larycheva Y 1996 Deformation design model of reinforced concrete elements under the action of bending moments and longitudinal forces *Concrete and reinforced concrete* **5** pp16-18

[4] Zvezdov A, Zalesov A, Mukhamedyev T, Chystiakov E 2002 Calculation of the strength of reinforced concrete structures under bending moments in longitudinal forces according to new regulations *Concrete and reinforced concrete* **2** pp 21-25

[5] Holyshev A, Bachynskyi V, Polyshchuk V i dr 2001 *Reinforced concrete structures* **1** Logos p 417

[6] Holyshev O B, Bambura A M 2004 *Lectures on the basics of calculating building structures and the resistance of reinforced concrete* Logos p 310

[7] EN 1992-1.1 Eurocode 2 *Design of concrete structures* Part 1-1 General rules and rules for buildings

[8] EN 1992-1-2 Eurocode 2 *Design of concrete structures* Part 1-2 General rules - Structural fire design

[9] EN 1992-2 Eurocode 2 *Design of concrete structures* Part 1-1 General rules and rules for buildings

[10] Mytasov V, Chkhum A 2018 Deformation of reinforced concrete beams with form-building elements under long-term loads. *Modern high technology* **3** pp 79-84

[11] Holodnov K A, Kondratiuk E V 2012 Stress-strain state of steel-reinforced concrete beams under various loads *Collection of scientific papers of the Ukrainian Institute of Steel Structures named after V. M. Shimanovsky* **10** pp 242-249

[12] Fan J, Nie J, Li Q, and Wang H 2010 Long-Term Behavior of Composite Beams under Positive and Negative Bending Experimental Study *Journal of Structural Engineering* **136(7)** pp 849–857. doi:10.1061/(asce)st.1943-541x.0000175

[13] Tan K H, and Saha M K 2006 Long-Term Deflections of Reinforced Concrete Beams Externally Bonded with FRP System *Journal of Composites for Construction* **10(6)** pp 474–482. doi:10.1061/(asce)1090-0268(2006)10:6(474)

[14] Gao D, Gu Z and Wu C 2020 Bending Behavior and Deflection Prediction of High-Strength SFRC Beams under Fatigue Loading *Journal of Materials Research and Technology* doi:10.1016/j.jmrt.2020.04.017

[15] Blikharskyi Z Ia 2005 *Stress-strain state of reinforced concrete structures in an aggressive environment under the action of loads* Dis ... doct. tech. Sciences 05.23.01 Lviv p 357
[16] Blikharskyi Z Ia, Vashkevych R V, Struk R F 2004 Methodology for calculating reinforced concrete structures recovered after corrosion *Bulletin of the Odessa State Academy of Civil Engineering and Architecture Odessa* pp 41-47
[17] Berh O Ia, Sherbakov E N, Pysarenko H N 1971 *High-strength concrete* Moscow Stroyizdat p 207
[18] Bazhenov V A, Huliar O I, Rutkovskyi V A 2003 Methods for numerical modeling of nonlinear deformation and fracture of spatial reinforced concrete structures *Collection of scientific papers Building structures* **59** pp 157-162
[19] Bambura A M 2006 *Experimental foundations of the applied deformation theory of reinforced concrete* Kharkiv p 39

Generalized optimality criteria of energy-efficient composites

V Kersh[1,2], **A Kolesnikov**[1,3], **N Khlytsov**[1,4] **and A Foshch**[1,2]
[1] Odessa State Academy of Civil Engineering and Architecture, 65029 Didrikhsona st., 4, Odessa, Ukraine
[2] Department of Urban Construction and Management
[3] Department of Chemistry and Ecology
[4] Department of Processes and Machine in Technology of Building Materials

E-mail: vkersh@hotmail.com

Abstract. The article discusses generalized optimality criteria of energy-efficient materials. The task of choosing an optimality criteria of a composite is considered as a step-by-step, hierarchical transition from individual criteria of the first level - operational characteristics to combined ones, related to subsequent levels and then to the general one. At each hierarchical level, the criteria become more complicated as it considers the criteria of the previous levels. The criteria constructed in this way are classified into structural, relative and conditional. It is shown that the softening coefficient and the water resistance index correspond to the criteria of the second and third levels, respectively. Some of the new combined criteria, having the character of efficiency were reviewed. As an example of a combined criteria, the criterion of the effectiveness of the realization of the strength-thermal conductivity ratio was considered. Examples of its use for evaluating and comparing the effectiveness of materials with significantly different properties and areas of use were given. It is shown that the efficiency criterion has similar values for materials with equal properties combination. On the basis of the proposed criterion, the problem of optimizing the efficiency of the plaster material is considered. As a result of numerical optimization, compositions with the maximum efficiency of the considered type were obtained. The possibility of further application of this and other combined optimality criteria in the problems of quality assessment and optimization of composite materials was shown.

1. Introduction

The main modern trend in the area of building materials science is the creation of multifunctional materials that simultaneously satisfy a set of requirements: energy-saving properties, high strength, adhesion and sound insulation, optimal environmental and economic characteristics. The current stage of the research development in the area of building materials science is associated with the development of multicomponent composite materials, which presupposes a preliminary assessment of their quality and a compromise optimization of their properties [1]. Due to this trend, it seems rational to consistently complicate the necessary optimality criteria of building composites.

2. The purpose of the research

This process can be based on a sequential grouping of individual characteristics, which ultimately leads to a single generalized criterion fig.1. In this case, the characteristics usually used for optimization (for example, thermal conductivity λ, compressive strength Rc and bending strength Rb, density d) can be represented as first level individual criteria.

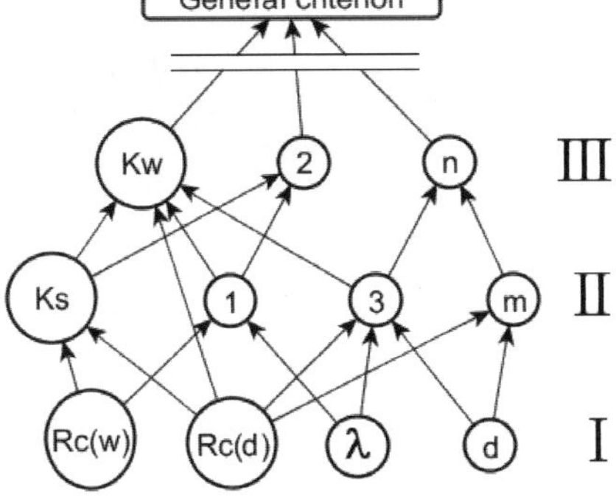

Figure 1. Hierarchy of optimality criteria for composite materials. I, II, III – hierarchical levels of criteria, Ks – softening coefficient, Kw – water resistance index, Rc(w) and Rc(d) – strength in water-saturated and dry states, respectively.

At each hierarchical level, the criteria are complicated by taking into account the criteria of the previous levels. Such complicated criteria of II, III and higher levels will be called combined. In this case, the combined criteria in a number of cases can be of independent interest and be useful for the research and optimization of materials of different classes.

3. Methods and Results of the Research

The most reasonable is the use of expressions in the form of products of criteria degrees of the previous level in a normalized form $\overline{C_i} = \frac{C_i}{C_{\max}}$ (normalization is made for the maximum value of a property in a group or for an interval of property change $\overline{C_i} = \frac{C_i - C_{\min}}{C_{\max} - C_{\min}}$). Combinations of criteria are carried out according to the rule:

$$C_{comb} = \prod_i \overline{C}_i^{n_i} \tag{1}$$

This introduction of combined criteria ensures its maximum value while simultaneously fulfilling the optimality conditions for all the criteria of the previous level. For example, the combined criterion - the water resistance index is equal to the product of the water-saturated strength (one criterion) by the softening coefficient (the second criterion) [2]:

$$\hat{E}_w = \frac{(R_{w.})^2}{R_d} = R_w \cdot K_p \tag{2}$$

The use of this combined criterion is quite effective in optimizing heat and sound insulating building composites with increased water resistance.
The proposed combined criteria can be classified as follows.

3.1. Structural criteria

They describe the relation between the structure and properties of the material, indicate the effectiveness of the implementation of the specified properties on the basis of the structure of the material organized using the control of recipe-technological factors. They have a general form $C_{SP} = P^n S^{-1}$, where P – desired property, n=1,-1, S – a measure of the number of structural units per volume, in the simplest case - density d. Examples of structural criteria:

IOP Conf. Series: Materials Science and Engineering 1162 (2021) 012005 doi:10.1088/1757-899X/1162/1/012005

a) $C_{d\lambda} = \frac{1}{d\lambda}$ – one of the criteria for the thermal insulation capacity of a material. Great for lightweight heat-insulating composites. By varying the structure of the material, one can solve the problems of structural optimization and, in particular, the problem of a conditional maximum $C_{d\lambda}$ – - to search the least dense material with a given thermal conductivity λ, or to minimize the thermal conductivity at a known density d.

b) $C_{dR} = \frac{R}{d}$ – effective strength criteria (bending, compression, adhesive). Shows how effectively the structure of the composite is organized for the implementation of the specified strength characteristics. Great for lightweight but durable materials. Such criteria can be used to solve optimization problems similar to those discussed above.

3.2. Relative criteria
Shows how much the desired property will increase with decreasing undesirable property by one, and vice versa. They are effective when taking into account the properties that change symbatically (for example, strength R – thermal conductivity λ). General view $C_R = \frac{P_d}{P_{ud}}$ – ratio of indicators of the desired property P_d to the indicator of undesirable P_{ud}. Example: $C_{R\lambda} = \frac{R}{\lambda}$ – efficiency indicator of the strength - thermal conductivity ratio. Great for energy-saving materials with increased strength. Since both of these properties are structurally determined, this criterion is combined with respect to structural.

3.3. Conditional criteria
Determine the degree of change in the properties of the material with a change in the technological factors of its manufacture or in case of extreme exposure (humidification, freezing, high temperature). This is the ratio of the conditional (for example, in extreme conditions) characteristics to the initial $C_c = \frac{P_c}{P}(\times 100\%)$; great for materials that are resistant to impact. One of the used criteria of this type is the softening coefficient $K_p = \frac{R_w}{R_d}$, equal to the ratio of the strengths of the material in the water-saturated state R_w and dry state R_d. So, the softening coefficient for dry gypsum $K_p = 0{,}4 - 0{,}5$, and in gypsum composites with loose hygroscopic fillers (for example, perlite, vermiculite) - even less, up to spontaneous destruction in a wet state. Creation of multicomponent materials based on gypsum (for example, gypsum-cement-slag binders - GCSB) and the use of hydrophobization makes it possible to obtain waterproof materials based on gypsum ($\hat{E}_{\eth} = 0{,}7 - 0{,}8$).

The degree of usefulness of the combined optimality criteria is determined, first of all, by the possibilities and results of their application in the practice of building materials science. The advantages of using them are as follows: criteria of the second and higher levels (fig.1) make it possible to reduce several operational properties to one value.

These values can be used to compare materials from different areas of application that have significantly different properties, but similar (comparable) values of the corresponding criteria. So materials with different sets of performance properties can be combined due to similar values of such criteria. Their general nature and interpretation can be made on the basis of the concept of the implementation effectiveness of the given relations. Example could be $C_{R\lambda}$. Consider the implementation of the comparison of the effectiveness of this type for composites of different nature and purpose. Compare efficiencies of 4 different composite materials type - gypsum perlite plaster composition [3] (hereinafter - the material samples 1, 18), heat-insulating polystyrene-gypsum concrete [4] (hereinafter - the material samples 2, 15) heat-insulating composition based on gypsum concrete [5] (hereinafter - the material samples 3, 15), lightweight spherogypsum concrete for tongue-and-groove slabs [6] (hereinafter - the material samples 4, 18). Their compressive strength Rc and thermal conductivity (Lambda), measured by normative methods, are shown in fig.2.

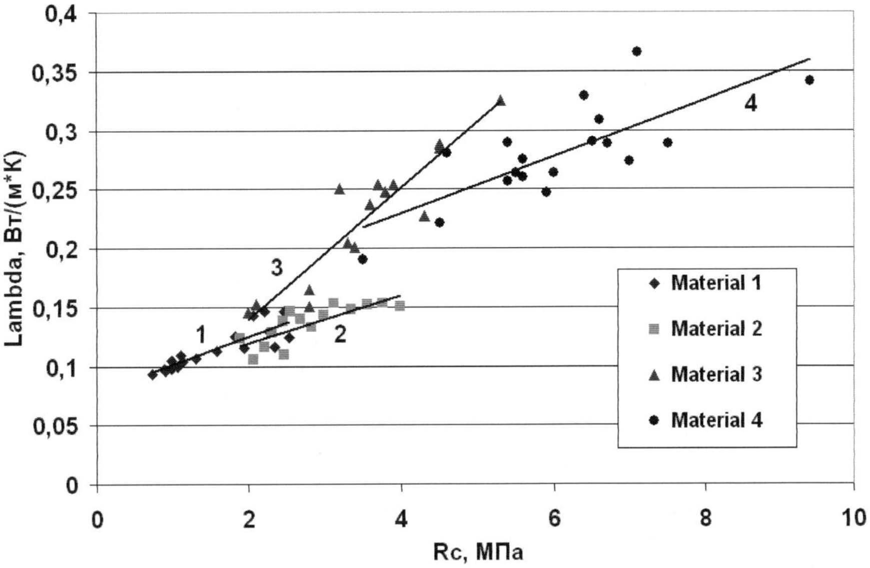

Figure 2. Thermal conductivity and compressive strength of materials 1-4

Figure 2 shows that materials 1 and 2 are characterized by similarity in the properties under consideration, while materials 3 and 4 differ significantly from the first group and differ from each other. Statistical researches of the combined group of materials are difficult. One option is to use an effectiveness parameter $C_{R\lambda}$. The values of this parameter for different samples of the corresponding materials 1–4 are shown in fig.3.

For materials substantially different in their initial properties Rc, Lambda of the materials in fig.3 regions of intersection of values $C_{R\lambda}$ are observed. Some samples from series 1–4 have similar values $C_{R\lambda}$. If we use $C_{R\lambda}$ to evaluate materials only by two properties, then the efficiency of materials 2 and 4 turns out to be higher than 1 and 3.

From the point of view of practical application, similar values of the considered efficiency and other similar criteria combine materials with the same efficiency. The choice of the material to use from such a group is carried out according to other characteristics (sound absorption, cost, etc.).

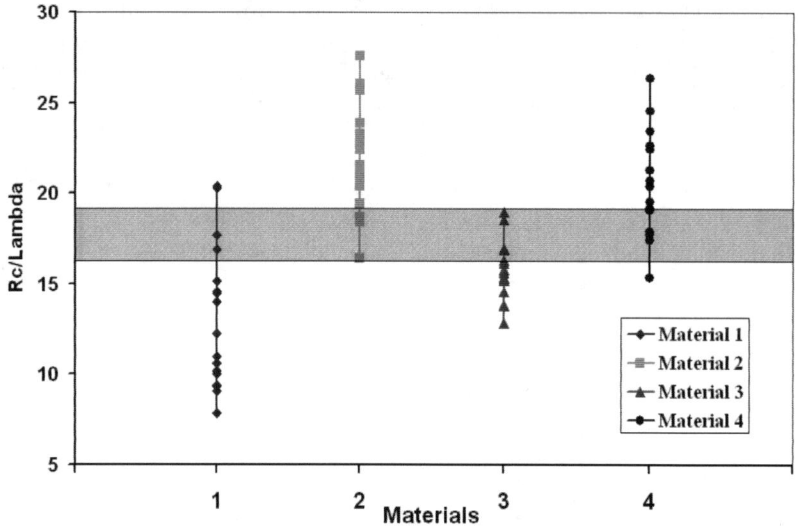

Figure 3. Comparison of values $C_{R\lambda}$ for composite materials1–4.
The intersection area $C_{R\lambda}$ for different materials is highlighted

Another way to apply the combined criteria is to use them as an indicator of the optimality of composites. They are easily calculated on the basis of already existing experimental data; regression (experimental-statistical) models can be built for them.

The criterion under consideration can, in particular, be used in experimental and statistical modeling of gypsum-perlite heat and sound insulating composition [3].

To study the properties and optimization of the material, a four-factor optimal experimental plan was used, the main factors and the levels of their variation were determined and shown in Table 1.

Table 1. Main factors and levels of their variation

i	Factors and levels of variation	$X_i = -1$	$X_i = 0$	$X_i = +1$
1	Perlite sand amount (volumetric parts per 1 part of gypsum), P/G	10	15	20
2	The amount of microspheres (Mks) and metakaolin (Mk) in their mixture (5% of the gypsum volume), MksMk	5% Mks	по 2.5% Mks and Mk	5% Mk
3	Superplasticizer dosage (% of gypsum volume), Plast	0.5	0.75	1.0
4	Latex content (% to the volume of the liquid mixture), Lat	1	1.5	2

On the basis of the values of compressive strength Rc and thermal conductivity λ (Lambda) obtained for 18 samples, the values of $C_{R\lambda}$ (Rc / Lambda) were calculated. For this characteristic an experimental-statistical (ES) model was built [7] fig.4.

IOP Conf. Series: Materials Science and Engineering 1162 (2021) 012005 doi:10.1088/1757-899X/1162/1/012005

Rc/Lambda =	+1.152 * MksMk * Plast	
+9.949	+1.653 * P/G2	
-3.550 * P/G	+2.510 * MksMk2	
+0.473 * MksMk	R-Squared	0.933
+0.392 * Plast	Adj R-Squared	0.886
+0.850 * Lat	Pred R-Squared	0.819

Figure 4. ES -model of the efficiency indicator and its statistical characteristics

The resulting model can be displayed as follows (fig.5.)

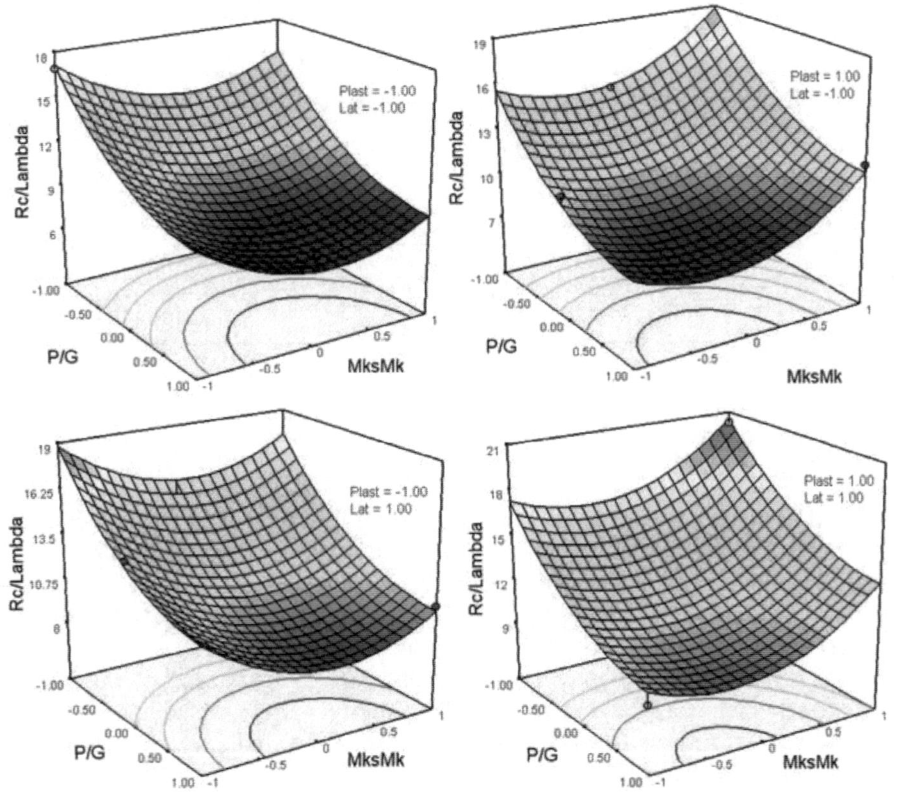

Figure 5. Graphical representation of the efficiency factor model $C_{R\lambda}$

ES-model of efficiency (fig. 4,5) allows to specify the values of the factors, necessary to achieve maximum efficiency – P/G = -1, MksMk = 1, Plast = 1, Lat =1. As the perlite-gypsum ratio decreases, the efficiency increases. It is characterized by the maximum values at the boundary points of the factor MksMk (1, -1), the ratio of the efficiency levels at the boundary points varies depending on the factors Plast, Lat.

The obtained efficiency criterion can be used in the numerical optimization procedure. The following structure of optimization tasks is proposed: individual operational properties (the first level criteria, fig. 1 are set within the framework of regulatory constraints, and the performance criteria are maximized. The optimization goals and the corresponding boundary values and importance are shown in Table 2.

Table 2. Optimization goals and parameters

Properties			Goal	Min.	Max.
d, кg/m^3	Density		in interval	450	600
Rb, MPa	Bending strength		in interval	0,5	1
Rc, Mpa	Compressive strength		in interval	1	2,5
Lambda, W/(m·K)	Thermal conductivity		in interval	0,1	0,2
Adhes, MPa	Adhesion strength		in interval	0,2	0,85
Sound*	Sound permeability		in interval	0,3	0,5
Rc/Lambda	Efficiency		maximum	8	20

* Sound permeability is a value equal to the ratio of the energy of the transmitted radiation to the energy of the incident radiation, $\frac{E_{ïðîø}}{\mathring{A}_{ïàä}}$. It shows the relative decrease in the force of sound when it passes through the thickness of the building material.

As a result of efficiency optimization, the compositions were obtained (table.3.)

Table 3. The most effective plastering compositions

Components				Properties						
P/G	MksMk	Plast	Lat	d	Rb	Rc	Lambda	Adhes	Sound	Rc/Lambda
-0,96	1,00	1,00	1,00	600	0,94	2,17	0,12	0,70	0,45	20,24
-0,95	0,93	1,00	1,00	600	0,93	2,13	0,12	0,68	0,44	19,76
-0,94	1,00	0,62	1,00	599	0,95	2,17	0,12	0,70	0,46	19,58
-0,89	0,97	1,00	1,00	593	0,91	2,06	0,12	0,68	0,43	19,56

4. Conclusions

Thus, the proposed strategy of using combined (generalized) optimality criteria of energy-efficient composites makes it possible to assess the effectiveness of multicomponent materials of increasing complexity based on individual operational characteristics. These performance criteria are useful for comparing materials with different sets of operational properties. "Equivalent" materials are characterized by similar values of the combined criteria. Some of them (water resistance index K_w, “strength-thermal conductivity” efficiency) can serve as the main part of the objective function for optimizing the structure of composites. For them, as well as for conventional operational characteristics, it is possible to build ES models. Application of optimization according to combined criteria leads to a decrease in the proportion of arbitrarily introduced auxiliary information of a subjective nature, which, in particular, improves the reproducibility of the optimization of the compositions of composite materials.

Thus, the development of combined optimality criteria seems to be a promising direction in building materials science, leading to the improvement of design processes, optimization and evaluation of the effectiveness of composite materials.

References

[1] Voznesenskiy V, Vyrovoy V and Kersh V 1983 *Modern methods of optimization of composite materials* Kiev: Budivelnik p 144

[2] Kersh V, Foshch A and Kolesnikov A 2017 Description of the water resistance of gypsum-containing compositions using a combined criterion *Collection of scientific papers of the Ukrainian State University of Railway Transport* **168** pp 133-140

[3] Kersh V, Kolesnikov A and Kersh D 2014 Development of rational insulating gypsum-pearlite *compositions Scientific and technical collection of the State Enterprise "Ukrainian Research Institute of Building Materials and Products" (NIISMI)* **52** pp 86–92

[4] Kersh V, Foshch A and Pyshchev O 2014 Polystyrene gypsum concrete for the production of tongue-and-groove slabs *Urban planning and spatial planning: scientific and technical collection* **51** pp 640-644
[5] Kersh V, Kholdaeva M and Shtets A 2010 Mathematical modeling and optimization of properties of lightweight concrete with polystyrene aggregate *Building materials, products and sanitary equipment* **36** pp 17-21
[6] Kersh D, Lyashenko T 2013 Optimization of the composition of spherical gypsum concrete *Building materials, products and sanitary equipment* **48** pp 149-153
[7] Lyashenko T and Voznesenskiy V 2017 *Methodology of recipe-technological fields in computer building materials science Training Manual* Odessa: Astroprint p 168

Variety of structure and changes in the properties of building composites over time

O Korobko [1], N Urazmanova [2], I Grynyova [3,5], L Dudnik [4]
[1, 2, 3, 4] Departament of Architectural structures; Odessa State Academy of Civil Engineering and Architecture, Didrikhsona st., 4, 65029, Odessa, Ukraine
[5] Address all correspondence to this author.

E-mail: grynyova@ogasa.org.ua

Abstract. The potential possibility of increasing the level of mechanical properties of composites on mineral binders remains during the long period. The initial structures hereditarily determine the change of the properties of the finished material in time. It can be assumed that an increase in the structural diversity of concretes should lead to such the organization of a structure that will guarantee the preservation or increase of the level of quality indicators at long terms of hardening (up to 720 days). The conducted researches showed that concrete of long hardening with the structure organized at selective adhesion of a cement matrix to the surface of dense or porous aggregates differed are more resistant to external influences. That is, those concretes, organization of structure which happened under the conditions of diversity of interphase interactions at the level of a macrostructure. This was ensured by the use of aggregates with various condition of a surface. The directed change of a ratios of adhesive-cohesive bonding forces at the surfaces of partition between a cement part and the surface of an aggregates provided structures, which caused increase the compressive strength of concrete by 18% and the module of elasticity by 24% at reducing the water by 29-36% and the open capillary pore volume by 44%. The structure of concrete organized at selective adhesion was the densest in comparison with of concrete of other structures. At the same time, the damage of material by technological cracks and inner surfaces of partition has grown more for concrete, the structure of which was provided by a variety of conditions of interactions between the cement matrix and aggregates. Favourable change of the parameters of these elements, along with the processes of hydration of relict binder grains, causes an improvement in time of indicators of the mechanical and deformative properties of concrete. Thus, structural diversity determines the potential changes of the concrete structure at long-hardening that must be taken into account at designing compositions and technology of the concrete, intended for difficult operating conditions.

1. Introduction

The concept of the "life cycle" of structures, buildings and structures is based on certain requirements. First, an assessment of the consumer value of construction projects should be carried out, depending on the provision of the required level of physical and technical properties of materials, taking into account the solution of the tasks of reducing their material consumption. Secondly, attention should be paid to the performance of the functions laid down in the material during the normalized operation time of structures and products. Maintaining the adjusted quality indicators of the cement and concrete component of structures and products during their long-term operation is especially important in case of probable changes in the intended purpose of construction projects. The maintenance of the design level of properties is provided by the structure that alone can guarantee the durability of the composites over time. For this, the structure must include a certain set of elements, the interaction of which is capable of determining an acceptable change in the physical and technical characteristics of the material of structures and products. The define role in the dynamics of the development of the entire structure, and, therefore, changes in properties affect cracks, internal interfaces and residual deformations as active elements of the structure, which instantly react to any intrastructural changes. The specific internal design of the structure, spontaneously organized by the beginning of the operation period, has the potential to perform certain actions aimed at self-preservation of the structure (product) as a single integral object without significant loss of functional features. Regulation of the variety of sets of components at different large-scale levels of organization of the structure makes it possible to obtain building composites with increased quality indicators. It is possible to purposefully create conditions for the formation of a certain "spectrum" of structures in the structure of one construct by increasing the variety of qualitative and quantitative compositions of binders. The prerequisites for the implementation of adequate behaviour of composites under changing conditions are set by the structure, which should ensure the preservation of the values of crack resistance and the level of mechanical characteristics over time. The authors proceeded from the assumption that an increase in the variety of parameters of active elements of the structure should have a positive effect on the change in the properties of materials over time. Determination of indicators of fracture toughness and strength of cement compositions after 8 years of hardening in natural conditions and assessment of the influence of a variety of structures of different ages on the change in the properties of composites over time was designated as a research task.

2. Literature review

The conducted review of scientific and technical information showed that at the moment there is no unified theoretically substantiated way for assessing the increase in the strength properties of concrete. Calculation of changes in concrete strength over time can be carried out in several ways: - according to the logarithmic law; - using generalized experimental strength growth factors; - based on the determination of the effect of an increase in the activity of the cement component or a decrease in the water-cement ratio on the growth rate of concrete strength through its nominal grade strength. Traditionally, most of the research is devoted to addressing the issues of improving the mechanical characteristics of building composites at a relatively early age [1, 2. 3, 4, 5]. Taking into account that the most significant investigation on the researched the strength of cement compositions and concretes in the late periods of hardening were carried out several decades ago, there are certain restrictions on the use of their results at the present time. The previously established quantitative ratios of concrete strengths at different curing times aren't rational to use in calculations without experimental verification and adjustment in accordance with the characteristics of modern cements and concrete production technology. Thus, an important problem is the accumulation of experimental results on determining the strength of cement stone, as well as concretes on dense and porous aggregates, over time.

3. Statement of the purpose and objectives of research

The aim of the research is to increase the level of properties of cement stone and concrete over time by increasing the diversity of the potential of their initial structure by introducing polydisperse fillers and changing the ratio of adhesion-cohesive bond forces at the interfaces between the cement matrix and the surface of fillers in concrete of the same composition. Research objectives: to determine the indicators of fracture viscosity and strength of cement compositions after 8 years of hardening in natural conditions and to assess the effect of a variety of structures of different ages on the change in the properties of composites over time; analyse the conditions for the initial structure formation of concrete at the macrolevel and determine the change in the properties of heavy concrete and expanded clay concrete after long hardening.

4. Methods of research

Cement compositions, including quartz fillers of rational qualitative and quantitative compositions, and concretes on granite crushed stone and expanded clay gravel were accepted as objects of research. For the manufacture of samples, cement was used as a binder, obtained by joint grinding in a laboratory ball mill of clinker manufactured by Closed Joint-Stock Company "Odessacement" and two-water gypsum stone in the amount of 5% of the cement mass.

To control the organization of the structure of cement compositions in order to increase the diversity of its characteristic components (clusters, initial cracks and internal interfaces, residual deformation fields), the amount of fillers N = 20 ± 10% of the volume of cement and their specific surface area S = 300 ± 200 m^2/kg. Fillers were introduced in various mono- and polydisperse ratios.

The experiments were carried out on standard beams 40x40x160 mm in size. The following properties of the samples were controlled: technological (initial) damage through the damage coefficient (K_d, cm/cm) [2]; compressive strength and flexural tensile strength (R_c, MPa, R_f, MPa) according to the standard method; crack resistance with the recommended methods of crack initiation by means of stress intensity factors ($K_{1s}{}^z$, $K_{1s}{}^r$).

The assessment of the influence of the diversity of the structure of different ages on the change in the properties of the composites was determined as the ratio F = Q^8 / $Q^{0.5}$, where: Δ F(Q) is the change in the property index of the cement stone over time; Q^{8years} - an indicator of the properties of a cement stone at the age of 8 years; Q^{28days} is an indicator of the properties of a cement stone at the age of 05 years (200 days) from the moment of hardening.

The study of the properties of concrete was carried out on samples-cubes made from a concrete mixture on dense and porous aggregates with different surface conditions. Together with traditional compositions of heavy and light concrete, compositions were used that contained aggregates that were fully or partially processed by GKR-11. Grains of crushed granite and expanded clay gravel were subjected to hydrophobization by immersion in a water repellent solution at a concentration of 2% from weight, followed by drying at T100C. This made it possible to provide different conditions for the interaction of the cement matrix with the surface of the aggregates, even with the same concrete composition: R_A>R_K, the adhesion of the matrix material to the aggregates is greater than its cohesive strength (concrete of composition I); R_A<R_K, there is no adhesion of the matrix component to the surface of the aggregates (concrete of composition II); R_A = R_K, volumes with perfect and imperfect adhesion of the matrix material to the surface of the aggregates (concrete of composition III) coexist in the material. R_A – is the adhesion of the matrix to the surface of the aggregates, R_K – is the value of the cohesive strength of the matrix material.

During the constant mobility of the concrete mixture, which required adjusting the compositions according to the values of the water-cement ratio, concretes with the corresponding property indicators were obtained. The change in the following properties of concrete and expanded clay concrete over time was monitored: average density (ρ_ω, kg / m^3), volume of open capillary pores (P_o,%), water absorption (W,%), velocity of ultrasonic waves (U, m/c), technological damage (K_d, cm / cm2), modulus of elasticity (E, MPa10^{-3}) and compressive strength ($f_{ck.cube}$, MPa). Property indicators were determined using standard methods.

The damage was quantitated using damage coefficients:

- as the ratio $K_d = \sum L_i / S$, cm / cm^2, where: $\sum L_i$ - is the total length of technological cracks and internal interfaces on the selected surface of the sample, cm; S - fixed surface area of the sample, cm^2;
- as the ratio $K_{d1} = \sum L_i / L_i$, cm / cm, where: $\sum L_i$ - the actual length of the fracture crack, cm; L_i - is the shortest distance between the points of exit of the actual fracture crack on the end surfaces of the sample, cm.

The impact of the age of concrete samples on the change in the level of their properties (Q) was assessed according to the corresponding coefficients: $\Delta\rho_\omega$, ΔPo, ΔW, ΔU, ΔK, ΔE, ΔF, which were determined as: $\Delta Q = f_{ck.cube}{}^{720} / f_{ck.cube}{}^{28}$, where: ΔQ – change an indicator of a specific property of concrete; Q^{720}– indicator of concrete properties at the age of 720 days; Q^{28} – is an indicator of concrete properties after 28 days of normal hardening.

5. The results of research.

Preservation of a preassigned level of properties it through time can be ensured by intensive changes in the structure, the adequacy of the manifestation of which depends on the possibility of realizing the interrelationships of all its individual components. This, in turn, is determined by the internal potential of the structure, which largely depends on the diversity of the set of its constituent parts.

The polymineralality and polydispersity of cement compositions a priori leads to the genesis of cluster structures with a unique combination of particles in them in terms of mineralogical composition, size

and quantity. The processes of multi-scale clustering and the emergence of "structures in the structure" interacting with each other to form even more hierarchically complex structures suggest a potential variety of parameters of initial cracks, interfaces, residual stresses and deformations. These integral elements of the structure of any composite material determine its potential for timely adaptation to internal and external influences. Taking into account the possible change in the operating conditions of building objects, it is necessary to provide for situations requiring the appropriate resistance of the material of products and structures. The prerequisites for the implementation of adequate behaviour of composites under changing conditions are set by the structure, which should ensure the preservation of the values of crack resistance and the level of mechanical characteristics over time. In this case, they proceeded from the assumption that an increase in the variety of parameters of active elements of the structure should have a positive effect on the change in the properties of materials over time.
The research results are presented in Table 1 and Table 2.

Table 1. Assessment of changes in the properties of cement compositions with monodispersed fillers over time

Indicator $\Delta F=Q^{8year}/Q^{28day}$	The compositions of fillers								
	S_1=100 m²/kg			S_2=300 m²/kg			S_3=500 m²/kg		
	10%	20%	30%	10%	20%	30%	10%	20%	30%
$\Delta F(P)$	0,97	1,00	1,03	1,00	0,99	1,04	0,97	1,03	1,07
$\Delta F(Rc)$	1,36	1,06	1,41	2,57	1,43	1,35	1,98	1,83	1,40
$\Delta F(R_f)$	1,01	1,37	1,25	1,09	1,16	1,17	1,11	1,00	0,98
$\Delta F(K1S^Z)$	0,52	0,80	0,21	0,36	0,31	0,78	0,38	0,49	0,44
$\Delta F(K1S^R)$	0,76	0,63	0,45	0,41	0,35	0,50	0,80	0,51	0,61

Table 2. Assessment of changes in the properties of cement compositions with polydisperse fillers over time

Indicator $\Delta F=Q^{8year}/Q^{28day}$	The compositions of fillers					
	10%		20%			30%
	$50\%S_1 + 50\%S_3$	$50\%S_2 + 50\%S_3$	$50\%S_1 + 50\%S_3$	$50\%S_2 + 50\%S_3$	$30\%S_1 + 30\%S_2 + 30\%S_3$	$50\%S_1 + 50\%S_2$
$\Delta F(P)$	1,04	1,06	1,01	1,03	1,05	1,04
$\Delta F(Rc)$	2,20	1,81	1,76	1,59	2,04	1,86
$\Delta F(R_f)$	1,79	1,59	1,79	1,24	1,14	1,30
$\Delta F(K1S^Z)$	0,48	0,49	0,94	0,51	0,89	0,49
$\Delta F(K1S^R)$	0,73	0,82	0,80	0,52	0,98	0,72

Cracks and internal interfaces are elements of the structure that arise at the initial stages of its organization and hardening. This presupposes the participation of cracks and interfaces in all further processes of the structural development of the material for many years. The presence of cracks and interfaces causes an uneven distribution of various deformations and stresses that permanently arise in the cement stone, due to the fact that the processes of hydration and the processes of internal mass transfer do not stop in it. A constant increase in density fluctuations and a permanently redistribution of local residual deformation fields was provided. This maintains the structure of the material in a non-equilibrium state. Permanent change in the structure of the cement stone will determine its damage by technological cracks and internal interfaces, fracture toughness, compressive and bending strength.

This maintains the structure of the material in a non-equilibrium state. Permanent change in the structure of the cement stone will determine its damage by technological cracks and internal interfaces, fracture toughness, compressive and bending strength. This manifests itself as structural changes in the material due to the fact that all the constituent structures are interconnected and mutually affect the state of each other. These are determined by the reorganization of the integral structure of the cement stone. The use of polydisperse fillers leads to a potential increase in the diversity of the material structure, regardless of the accepted ratios of sizes and number of particles. This, as shown by researchers to determine the damage of cement samples, leads to an increase in K_d values during prolonged hardening under normal conditions. Damage to samples with monodisperse fillers increased only at certain compositions, or practically did not change. The interaction of filler particles with the multivariance of the ratios of their geometric parameters led to the organization of primary cluster structures, which provided a more varied set of initial cracks and internal interfaces, which influenced the changes in the structure of the cement stone over time.

Technological damage determines the cracking resistance of materials. Analysis of the results obtained for the determination of K_{1s}^{z} and K_{1s}^{r} showed a general tendency towards a decrease in the failure viscosity of cement stone during long-term hardening: by 24-41% for samples with polydisperse fillers and by 44-53% for samples with monodisperse fillers. At the same time, during using filler compositions with a specific surface area of S_1+S_2 or S_1+S_3 in a 1:1 ratio, it is possible to increase the cracking resistance of a cement stone to 11% and 20%, respectively. After the introduction of fillers with the composition 33%S_1+33%S_2+33%S_3, the failure viscosity over time of the hardened cement compositions practically didn't change.

Strength indicators increase with time for all the accepted compositions of polydisperse fillers, up to 31% (R_f) and 47% (R_c). The structures formed upon the introduction of monodisperse fillers ensured an increase in the strength of the samples in bending and compression, on average, up to 13% and 30%, respectively. In systems with a wide variety of structures, the effect of the accepted age of the samples on the change in compressive strength was up to 2 times higher than those of systems that included filler particles close in specific surface area. Thus, depending on the composition of the fillers, the parameters of the substructures are set at the level of interaction of the binder particles. Thus, the conditions of hydration are determined, which affects the provision of sets of cracks and internal interfaces, on which the potential ability of the cement stone to reorganize the structure over long time intervals depends. Favourable structural changes lead to an increase in the strength of the cement samples.

The potential structure of the concrete begins to organize with the final distribution of the initial components after the completion of the formation of the product. The limited manifestation of sedimentation phenomena in the compacted concrete mix makes possible a fixed distribution and mutual orientation of aggregates with the formation of certain interfaces between them and the matrix. Thus, the prerequisites are laid for the formation of characteristic elements of the material structure at the macrolevel in the form of peculiarly ordered cells. A more rational approach is to model concrete as a set of cells with different parameters. This model reflects the variety of concrete at the matrix material - aggregates level. All processes of interaction of the matrix material and aggregates depend on the shape of the cells and the ratio of the values of cohesion of the soluble part and its adhesion to the surface of grains of crushed stone or gravel. Products based on concrete of industrial compositions include various sets of cells in terms of geometric and physical parameters. In this case, the shape and size of the cells are spontaneous, however, a directed change in the state of the surface of the aggregates contributes to an increase in the variety of cells in concrete of the same composition.

Due to the different orientations of the fillers and different ratios of the bonding forces on the interfacial surfaces in each cell, there is an individual manifestation of the properties of the matrix material as a discrete medium [4]. Depending on the shape of the cells and the state of the surface of the aggregates, a unique block structure of the matrix component is formed. This, in turn, affects the length of the internal surfaces of the matrix interface in individual cells, which affects the manifestation of volumetric deformations and the setting time of the material as a result of changes in the conditions of the hydration reactions.

The researches carried out on physical models of structural cells made it possible to determine that the length of the external boundaries and the area of the matrix material can change up to 39% and 35%, respectively, depending on the geometric characteristics of the cells. Different conditions of adhesion of the matrix to the surface of the fillers in the cells cause a change in the size of the matrix component by 27-36% and a change in the density of the material up to 40%. The initial volumetric deformations of the matrix can differ in magnitude by 10-26%, in the kinetics by 7-44%. During transition from a cell of one form to cells of other forms, the setting time of the material changes by 20-150 minutes. The geometric parameters of the cells and the ratio of R_A and R_K values at the interfaces between the mortar component and aggregates are a powerful factor in controlling the structure formation of concrete at the micro level in the early periods. Thus, an increase in structural variety due to the directional obtaining of a complex set of initial structures at the level of binder particles by changing the parameters of the macrostructure provides the structural and functional potential in all material. This should affect the mechanical and deformative characteristics of concrete over time.

Table 3 shows the analysis of the coefficients obtained from the experimental results, the change in the level of properties of concrete on dense and porous aggregates in time was determined by the initial structures provided with different ratios of the binding forces at the interface of the cement matrix with the surface of the aggregates (R_A/R_K).

Table 3. Analysis of changes in the properties of concretes with different surface conditions of aggregates during long-term hardening

$R_A>R_K$ Ratios	Coefficients for assessing changes in concrete properties over time						
	$\Delta\rho_\omega$	ΔPo	ΔW	ΔU	ΔK	ΔE	ΔF
			Concrete on crushed granite				
$R_A>R_K$	1,01	0,66	0,69	1,12	1,05	1,23	1,03
$R_A<R_K$	0,96	0,70	0,74	1,12	1,04	1,17	1,13
$R_A=R_K$	0,95	0,55	0,64	1,17	1,07	1,32	1,20
			Concrete on expanded clay gravel				
$R_A>R_K$	0,96	1,0	0,87	1,0	1,01	0,97	1,10
$R_A<R_K$	0,96	0,82	0,78	1,03	1,03	1,0	1,15
$R_A=R_K$	0,98	0,89	0,71	1,0	1,10	1,01	1,22

Under the conditions of the experiment, concrete of composition I and expanded clay concrete of composition III were distinguished by a minimum decrease in average density indicators over time. However, all types of concrete with a structure organized at $R_A = R_K$ had the highest values at 28 and 720 days of hardening. The volume of open capillary pores in concretes of the accepted compositions decreased over time. The maximum Po indicator decreased with the composition of III concrete, up to 44%, and with the composition II of expanded clay concrete, up to 18%. After long-term hardening during 720 days, the lowest values of water absorption were observed for concrete and expanded clay concrete, the structure of which was initiated under conditions of selective adhesion of the matrix to the surface of the aggregates. Thus, concretes with $R_A = R_K$ are denser in the accepted time interval in comparison with concretes of other compositions. This is confirmed by the preservation at the initial level of the rates of ultrasound transmission in the samples of concrete and expanded clay concrete. At the same time, the damage caused by technological cracks and internal interfaces increases to a greater extent for concretes, the structure of which was provided by a variety of ratios of values R_A and R_K. Damage to heavy concrete increased during structure formation under conditions $R_A = R_K$ by 21% and 29%, compared to structures organized under $R_A<R_K$ and $R_A>R_K$ conditions, respectively. For expanded clay concrete samples, the selectivity of matrix adhesion to the surface of aggregates leads to the formation of structures that provide a damage value that is less than the values of K_p of expanded clay concrete of composition I - by 39%, composition II - by 19%. The development of the main crack repeats the trajectory of operational cracks and internal interfaces, which accumulate in the material under the action of various types of loads and arise as the growth of technological cracks present in the material for operation [5]. This assumes that the geometric parameters of cracks and fractures are determined by the number and relief of the banks of hereditary cracks and internal interfaces, oriented along the front of their movement. The value of the damage coefficient Kp1, through which it is possible to estimate the length of the fracture crack, with the structure initiated under the conditions of R_A and R_K, were 7-18% less than the indicators K_{p1} of the samples with the structure formed with perfect or absent adhesion of the matrix component to the surface of the aggregates.

Structural changes associated with the growth and transition of cracks to the rank of internal interfaces, along with the processes of hydration of relic binder grains, cause an improvement in the level of mechanical and deformative properties of concrete and expanded clay concrete over time. During selective adhesion of the matrix to the aggregates, structures are formed that provide the maximum increase in the elastic modulus of concrete under experimental conditions, up to 24%. The E values for expanded clay concrete are maintained at the original level during the long-term approval. After providing a variety of initial conditions for the interaction of the matrix material with aggregates, the compressive strength of concrete increases with time by 16%, the values of the compressive strength of expanded clay concrete increase by 18%. The greatest increase in strength under experimental conditions is observed in concrete and expanded clay concrete of composition III.

6. Conclusions

It can be concluded that increasing the diversity of the set of characteristic structural components by introducing rational ratios of mono- and polydisperse fillers and changing the ratio of adhesive-cohesive bond forces between the matrix component and aggregates allows maintaining the design level of mechanical properties and crack resistance of cement compositions and concretes over time. The change in the damage of the cement stone during long-term hardening in advantageous conditions indicates continuous structural transformations, the main role in which is played by technological cracks. An increase in various combinations of parameters of active elements of the structure determines the timeliness of their reactions to the changing dynamics of hydration processes of relict binder grains. This contributes to the maintenance of the specified quality indicators of the cement component during the long-term operation of building products and structures.

The use of a mixture of aggregates with different surface conditions led to the formation of structures in which the compressive strength of concrete increased to 18%, their modulus of elasticity increased to 24%, water absorption rates decreased to 29-36% with a decrease in open porosity to 44%, the length of cracks destruction decreased by 7-18% with an increase in technological damage by 19-39%. The change in damage and parameters of main cracks in time is determined by the organization of the structure of building composites at the early stages of structure formation. Thus, the initial structure under the influence of the accepted structure forming factors hereditarily determines the structural changes and the manifestation of the physical and technical properties of concrete and expanded clay concrete during prolonged hardening. It is recommended to take into account the results of the investigation carried out during prescribing the compositions of the cement binder and concrete mixtures in the manufacture of small-piece products and monolithic housing construction.

References

[1] Abdelmonem A, El-Feky M S, Nasr El-Sayed A R and Kohai M 2019 Performance of high strength concrete containing recycled rubber *Construction and Building Materials* **227** 116660

[2] Deng X-F; Liang Shi-Lin; Fu F and Qian K 2020 Effects of High-Strength Concrete on Progressive Collapse Resistance of Reinforced Concrete Frame *Journal of Structural Engineering* **146 (6)** pp 47-55

[3] Vishalakshi K, Revathi V and Sivamurthy S 2018 Effect of type of coarse aggregate on the strength properties and fracture energy of normal and high strength concrete *Engineering Fracture Mechanics* **194** pp 52-60

[4] Zhao H, Jiang K, Yang R, Tang Y and Liu J 2020 Experimental and theoretical analysis on coupled

effect of hydration, temperature and humidity in early-age cement-based materials *International Journal of Heat and Mass Transfer* **146** 118784
[5] Shi J, Liu B, He Z, Wu X, Tan J, Chen J, Jiang J 2020 Properties evolution of high-early-strength cement paste and interfacial transition zone during steam curing process *Construction and Building Materials* **252** 119095

Research of crack formation in inseparable double-span reinforced concrete elements

O Krantovska[1,2], L Ksonshkevych[1,3], M Petrov[4] and S Synii[5]

[1] Odessa State Academy of Civil Engineering and Architecture, 65029, Didrikhsona st., 4, Odessa, Ukraine
[2] Department of materials strength
[3] Department of Urban Construction and Management
[4] Odessa National Academy of Telecommunications named after O. Popov, 65029, Kovalska st. 1, Odessa, Ukraine
[5] Lutsk National Technical University, 43018, Lvivska st. 75, Lutsk, Ukraine

E-mail: elena14122007@gmail.com

Abstract. According to the results of the conducted, experimental researches empirical dependences of crack formation and width of opening of normal and inclined cracks of inseparable double-span reinforced concrete beams. A reliable quantitative and qualitative assessment of the impact of structural factors, as well as external factors on the crack resistance of inseparable elements, both separately and in interaction with each other. An improved deformation method (implemented in the PC MATLAB) for calculating the crack resistance of reinforced concrete elements. Based on the comparative analysis of the obtained empirical dependences and calculated values according to normative methods: DSTU B B.2.6-156: 2010 (Ukraine), previously acting SNiP 2.03.01-84 *, SR 63.13330.2012 (SET OF RULES, actualized) Russian standard, as well as the deformation model of Scientific Research Institute of Building Constructions (SRIBC), town of Kyiv, the coefficient of variation of crack resistance and width of opening of normal and inclined cracks is determined. The comparative analysis is carried out and informative graphic drawings are presented.

1. Introduction

The peculiarity of the work of inseparable double-span reinforced concrete beams is the redistribution of internal efforts in the sections, so the appearance of primary and secondary cracks, their development, the opening of normal and inclined cracks has a great impact on their performance. In the process of loading the experimental elements, the first normal cracks appear above the middle support in the area of maximum bending moment, and then or simultaneously - under the average forces and in the span of the stretch of the stretched area. This feature of crack formation is due to the work of beams as once statically indeterminate elements, in which the bending moment over the middle support is several times greater than the moments under concentrated forces.

The purpose of the research is to reveal the nature of crack formation, both normal and inclined cracks above the middle support and under the concentrated forces of inseparable double-span beams, to investigate the influence of structural factors.

Perform a comparative analysis of the experimental values of the studied parameters with the calculated ones performed according to normative methods and the improved deformation model.

A large number of scientific works are devoted to crack resistance, deformability and strength of reinforced concrete elements [1-5]. The strength of normal and inclined sections was studied by T.N. Azizov H.S. Aliev, Y.M. Babych, A.Y. Barashykov, L.V. Baykov, Z.Y. Blikharskii, O.O. Hvozdev, O.S. Horodetskii, B.H. Demchyna, S.A. Dmytriev, L.O. Doroshkevych, V.S. Dorofeev, O.S. Zaliesov, V.M. Karpiuk, P.M. Koval, R.L. Mailyan, A.M. Pavlikov, O.I. Storozhenko, O.V. Andriichuk, Y.O. Shkola, O.F. Yaremenko, M.M. Petrov etc. Among foreign works can be noted the work of R. Walter, A. Vista, L. Cany, M.F. Collins, F. Leongardt, J. Mitchell, K. Modi, J. Morrow, J. Regan, J. Taub, D. Hanson etc.

2. Research Methodology

2.1 Experimental research

In accordance with the developed methodology of experimental research have been tested two-span continuous beams, respectively, b×h =100×180 mm with h_0 =155 mm, l=8·h_0, a/h_0=1, 2, 3, made of heavy concrete of classes C12/15, C20/25, C30/35, reinforced by two plain frames with the lower rebar 2Ø10, 12, 14 500S, with the top rebar 2Ø10, 12, 14 A500S and with transverse rebar 2Ø3, 4, 5 Vr-I. Experiments were performed on five-factor, close to properties of the D-optimal plan of type H5, which provides the same accuracy of the prediction of the experimental value in the area described by a radius equal to 1 relative to the central "0" point (Experiment № 27).

2.2 The mechanism of cracking, deformation and fracture of a continuous double-span beam

The initial stage of crack development is characterized by the concentration of concrete deformations, especially at the level of the location of the stretched reinforcement above the middle support.

In the process and after the formation of a plastic hinge over the middle support, which is characterized by the fluidity of the upper longitudinal reinforcement of the beam and (or) crushing of the concrete of the compressed zone, there is a redistribution of internal forces with a sharp increase in bending moment under concentrated forces and holding its maximum value above the specified support, which confirmed by studies [3-7].

Before the appearance of normal cracks, the distribution of internal forces corresponds to the rules of structural mechanics. After the appearance of the first cracks above the middle support (stage I, Fig. 1), inelastic properties appear and the forces are found in the diagram σ-ε. Upon reaching the yield strength over the middle support, i.e. the formation of a plastic hinge, the beam is divided into two split with concentrated moments above the middle support (stage III, Fig. 1).

When bending a beam with a moderate content of transverse reinforcement in the zones of action of bending moments and transverse forces as the load increases and the development of inclined cracks, the beam is divided into parts connected by the concrete of the compressed zone and the reinforcement intersecting by cracks. The destruction of the beam is characterized by a sharp opening of one of the inclined cracks - the so-called critical with the following physical destruction of concrete over this crack due to the achievement of longitudinal deformations in the direction of action of the main compressive stresses of the limiting values that manifest themselves either as external crushing of concrete or as crushing with a puncture. The stress state of the compressed zone at the approach to fracture is complex, since shear stresses act next to the compressed ones.

IOP Conf. Series: Materials Science and Engineering 1162 (2021) 012007 doi:10.1088/1757-899X/1162/1/012007

Graph of bending moments M_x before the appearance of the «plastic hinge»

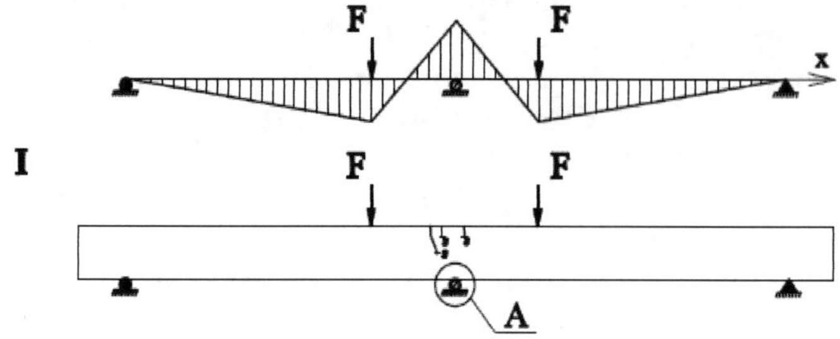

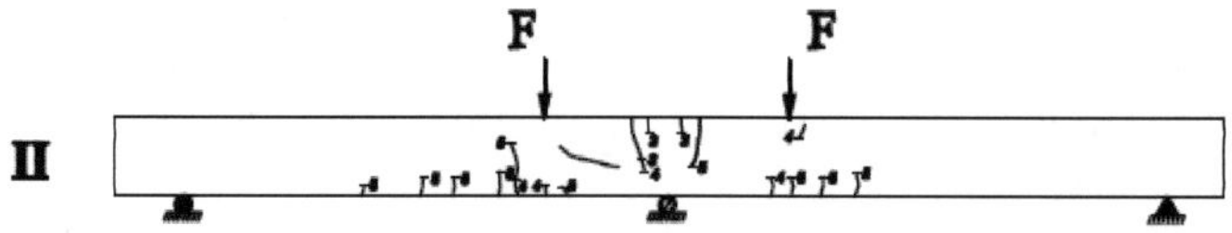

Graph of bending moments M_x after the appearance of the «plastic hinge» over the middle support

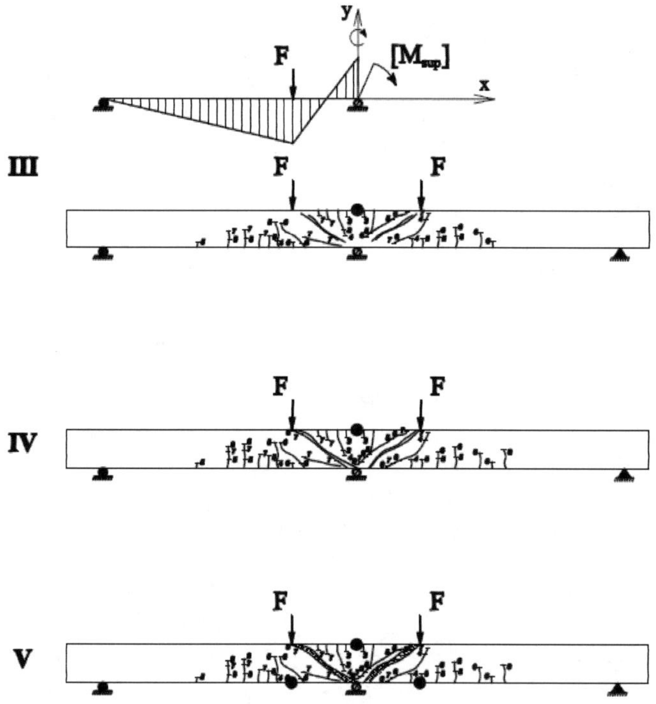

Figure 1. The mechanism of cracking, deformation and fracture of a continuous double-span beam. (I - the appearance of normal cracks above the middle support, node *A* - dynamometer; II - the development of normal cracks above the middle support, the appearance of normal cracks under the load and (or) inclined slots in the span; III - the development of normal and inclined cracks, the appearance of a "plastic hinge" above the middle support; IV - development of normal and inclined cracks with a predominant opening of inclined cracks, up to failure along inclined sections; V - destruction along inclined sections with the formation of a "plastic hinge" under concentrated loads in beams with large span sections).

3. Results

Processing of the experimental data presented in [6-9] on the resistance to cracking, deformation and fracture of integral double-span beams, allowed to obtain mathematical models (1-8) by the method [9], which can be used to assess the impact of relative shear span (experimental factor *X1*, a/h_0=1,2,3), concrete class (experimental factor *X2*, C, MPa: C12/15; C20/25; C30/35), number of transverse reinforcement (experimental factor *X3*: ρ_w (BpI) = 0.0018: 0.0032; 0.0050 (2Ø3.4.5 at S=7.75 cm)), the number of lower and upper longitudinal reinforcement (experimental factor *X4*: $\rho_{f,lower}$ (A500C) = 0.010; 0.0146 ; 0,0146 (2Ø10, 12, 14), *X5*: $\rho_{f,upp.}$ (A500C) = 0,010; 0,0146; 0,0146 (2Ø10, 12, 14) on their crack formation:

$$YV_{crc,\perp}^{mid.sup.} = 16.6 - 6.0X_1 + 4.6X_2 + 0.9X_5 + 3.7X_1^2 - 1.3X_1X_2,\ кH,$$

$$\Sigma = (Y - \bar{Y})^2 = 22.50;\quad \sigma = \sqrt{\frac{(Y - \bar{Y})^2}{n - 1}} = 0.9;\quad \upsilon = \frac{\sigma}{b_{0,res.}} \cdot 100\% = 5.6\%; \tag{1}$$

$$YM_{crc,\perp}^{mid.sup.} = 3.4 + 0.8X_2 + 0.16X_5,\ кHм,$$

$$\Sigma = 0.75;\quad \sigma = 0.17;\quad \upsilon = 5.0\%; \tag{2}$$

$$YV_{crc,\perp}^{und.forc.} = 62 - 38.8X_1 + 11.9X_2 + 14.4X_5 + 17.4X_1^2 - 9.4X_1X_5, кH,$$

$$\sigma = 6.4;\quad \upsilon = 10.4\%; \tag{3}$$

$$YM_{crc,\perp}^{und.forc.} = 3.4 + 0.8X_2 + 0.14X_4, кHм, \qquad \sigma = 0.2;\quad \upsilon = 5.5\%; \tag{4}$$

$$YV_{crc,/} = 40.9 + 9.6X_2, кH, \qquad \sigma = 2.3;\quad \upsilon = 5.6\%; \tag{5}$$

$$YM_{crc,/}^{mid.sup.} = 8.4 + 2.2X_1 + 1.8X_2 - 0.9X_1^2 + 0.6X_1X_2, кHм,$$

$$\sigma = 0.5;\quad \upsilon = 5.5\%; \tag{6}$$

$$YM_{crc,/}^{und.forc.} = 3.2 + 2.5X_1 + 0.8X_2 + 0.14X_1^2 + 0.6X_1X_2, кHм,$$

$$\sigma = 0.2;\quad \upsilon = 5.4\%; \tag{7}$$

$$YQ_{crc,/} = 37.4 - 3.4X_1 + 8.6X_2 - 0.7X_1^2 - 0.6X_1X_2, кH, \qquad \sigma = 2.0;\quad \upsilon = 5.4\%; \tag{8}$$

Adequate mathematical models of external load, which causes the appearance of normal cracks above the support (1) and under concentrated forces (3) in the rungs, indicate that the greatest influence is exerted by the relative shear of the cut, concrete class and the number of upper longitudinal reinforcement.

The values of the moments above the average support (2) and the concentrated forces (4), at which normal cracks are formed, increase in relation to their average value:

with an increase in the class of concrete from C12/15 to C30/35 - up to 48%;

with an increase in the longitudinal upper $\rho_{f,upp}$ (A500C) from 0.010 to 0.0146 (lower $\rho_{f,lower}$ from 0.010 to 0.0146) rebar, in accordance, - on 9.4% and 8.3%, that is, insignificant.

An analysis of the mathematical model of the transverse force at which inclined cracks appear (5) indicates that only the concrete class has the greatest influence. When it increases from minimum to maximum values $V_{crc,/}$ increases by 47% relative to its average value

The moments of occurrence of inclined cracks over the middle support (6) and average forces (7) are influenced by the relative run of the cut and the class of the concrete.

The transverse force corresponding to the appearance of inclined cracks (8) will increase in relation to their average values at:

reduction of the relative run of the slice a/h_0 from 3 to 1 - by 18%;

increased concrete grades from C12/15 to C30/35 - by 46%.

The presence of a negative sign of the quadratic effect X_1^2 indicates that with decreasing slice run, the experimental parameter will increase more rapidly.

Below are the results (Fig. 2) of the comparative analysis of the obtained empirical dependences of crack resistance, width of opening of normal and inclined cracks with calculated values according to the improved deformation technique, normative methods (National standard of Ukraine DSTU B V.2.6.-156: 2010, previously acting SNiP 2.03.01-84 *, SR 63.13330.2012 (SET OF RULES, actualized) Russian standard, as well as the deformation model of Scientific Research Institute of Building Constructions (SRIBC), town of Kyiv [11].

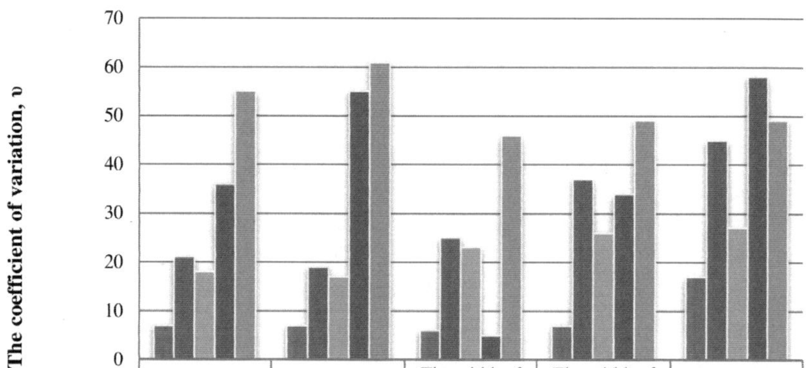

	The moment of appearance of normal cracks, Mcrc	The moment of appearance of inclined cracks, Mcrc,/	The width of the opening of normal cracks above the middle support, W	The width of the opening of normal cracks under concentrated forces, W	The width of the opening of inclined cracks, Wcrc/
■ Advanced deformation method (PC MATLAB)	7	7	6	7	17
■ Ukrainian standard DSTU B V.2.6-156:2010	21	19	25	37	45
■ Deformation method SRIBK (A.M. BAmbura)	18	17	23	26	27
■ Previously acting SNiP2.03.01- 84*	36	55	5	34	58
■ SR 63.13330.2012 (SET OF RULES, actualized) Russian standard	55	61	46	49	49

Figure 2. Comparison of the results of calculations of the coefficient of variation.

Among the important criteria chosen for the comparative analysis, the coefficient of variation in strength was selected (Fig. 2):

$$\upsilon = \frac{\sigma}{b_{0,res.}} \cdot 100\%, \quad (9)$$

where $\sigma = \sqrt{\sum \frac{(\hat{V}_{\exp} - V)^2}{n-1}}$; $\sum(\hat{V}_{\exp} - V)^2$ - is average square deviation; $b_{0,res.}$ - is arithmetic average experimental value of bearing capacity $\hat{V}_{\exp}$; n - number of test specimens.

4. Appendix

Photographic recording of experimental samples is shown below (Fig. 3)

a)

b)

c)

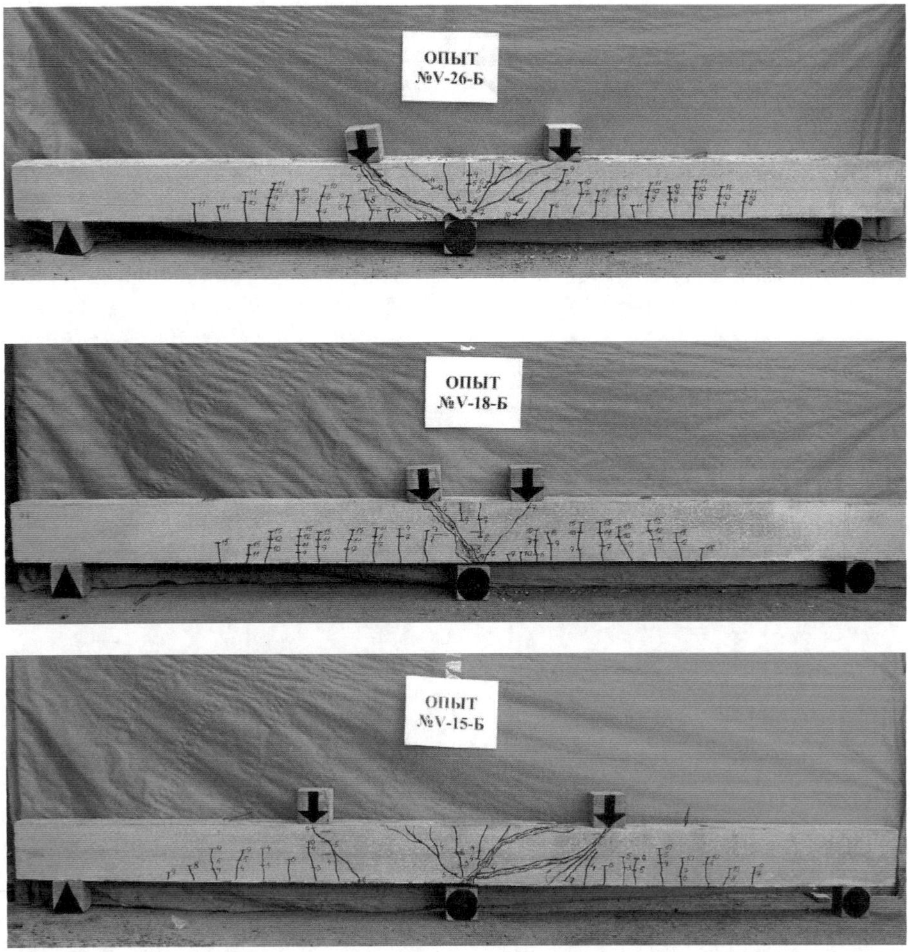

Figure 3. The nature of cracking and destruction of experimental two-span continuous beams with small (a), middle (b) and large (c) shear span.

5. Conclusions

Experimental-theoretical studies allowed to perform a reliable quantitative and qualitative assessment due to the action of structural factors, as well as factors of external action on the deformability, stability of inseparable elements, both separately and in interaction with each other. With a decrease in the run of the cut, the crack resistance of normal sections increases by 18%; with the increase of the concrete class, the crack resistance of normal and inclined increases to 48%, the amount of transverse, longitudinal lower and upper reinforcement is insignificant - up to 9.5%.

Comparison of experimental and calculated values $V_{crc,\perp}$, $V_{crc,/}$, $M_{crc,\perp}$, $M_{crc,/}$ above the middle support and under concentrated forces, as well as the width of normal and inclined forces $W_{\perp}$, $W_{/}$ showed that the use of deformation model SRIBC (Kyiv) [11] and the proposed deformation model using PC MATLAB, allows to achieve satisfactory convergence of experimental and calculated values of loads that correspond to the appearance of normal (υ =6.0…6.5%) and inclined cracks (υ=6.6…17%).

Comparison of experimental and calculated values of the opening width of normal and inclined cracks (Fig. 2) showed generally unsatisfactory convergence (υ to 58%).

References

[1] Babych E and Andriichuk O 2017 Strength of Elements with Annular Cross Sections Made of Steel-fiber-Reinforced Concrete Under One-Time *Materials Science* (New York) Vol. 52 Issue 4 pp 509-513

[2] Andriichuk O, Babich V, Yasyuk I and Uzhehov S 2017 The influence of repeated loading on work of the steel fiber concrete drainage trays and pipes on the roads *6th Int. Sc. Conf. Transbud-2017 (Kharkiv)* ed G Vatulia et al *MATEC Web Conf.*116 02001

[3] Babych E, Andriichuk O, Ninichuk M, Kysliuk D 2021 Experimental Research of Strength Characteristics of Continuous Reinforced Concrete Beams with Combined Reinforcement, and Modelling Their Work by the Finite Element Method *Lecture Notes in Civil Engineering* **100** *Springer, Cham*

[4] Soetens T, Matthys S 2014 Different methods to model the post-cracking behaviour of hooked-end steel fibre reinforced concrete *Constr. Build. Mater.* **73** p 458–471

[5] Kinash R, Bilozir V 2015 Deformational calculation method of bearing capability of fiber-concrete steel bending elements *Czasopismo Techniczne*

[6] Krantovska O, Ksonshkevych L, Petrov M, Synii S and Ksonshkevych S 2019 Deflections of continuous reinforced concrete elements IOP Conf. Series: *Materials Science and Engineering* **708** 012061 IOP Publishing

[7] Krantovska O, 2010 *Strength, crack resistance and deformability of continuous reinforced concrete beams: Dis. cand. tech. sc.* Odesa: OSACEA

[8] Krantovska O, Petrov M, Ksonshkevych L, Synii S, Sunak P 2018 Improved engineering method for calculating the strength of the supporting areas of reinforced concrete elements *Reliability and Durability of Railway Transport Engineering Structures and Buildings:* MATEC Web of Conferences, Kharkiv

[9] Krantovska O et al. Numerical simulation of the stress-strain state of complex-reinforced elements Tehnički glasnik, : https://doi.org/10.31803/tg-20190417112619

[10] Voznesensky V 1981 Static methods of experiment planning in technical and economic research *Finance and Statistics* Moscow

[11] Babaiev V, Bambura A, Pustovoytova O and others 2015 Practical calculation of elements of reinforced concrete structures on DBN.2.6-98:2009 in comparison with calculations according to SNiP 2.03.01-84* and EN 1992-1-1 (Eurocode 2), Kharkiv, Golden pages

IOP Conf. Series: Materials Science and Engineering 1162 (2021) 012008 doi:10.1088/1757-899X/1162/1/012008

Steel fibrous concrete with high-early strength for rigid pavements repair

S Kroviakov [1], V Kryzhanovskyi [1], M Zavoloka[1]

[1] Odessa State Academy of Civil Engineering and Architecture, 65029, Didrikhsona st., 4, 65029, Ukraine

E-mail: vitolloscience@gmail.com

Abstract. Experimental results, including the selection of repair concrete and steel-fiber-concrete mixtures for rigid pavement repair are presented. The complex effect of the superplasticizer, hardening accelerating agent and steel anchor fiber on the concrete mixture properties has been studied. Compressive and flexural strength characteristics of the repair material were determined in the early and design ages of hardening. Using the method of experimental statistical modeling, the optimal dosages of dispersed reinforcement in combination with accelerating agent and workability modifier have been determined.

1. Introduction

In world practice, roads with rigid pavement are actively used [1]. Ukraine is also increasing the pace of highways construction using rigid road pavements [2]. For maintenance, permanent and major repairs, as well as reconstruction of this pavement types, it is necessary to provide for special technological methods using high-early strength concrete compositions. The repair technology with asphalt concrete, generally accepted in Ukraine, does not justify itself in view of the material pliability under dynamic loads, especially at high temperature drops. Traffic density and axle load increase and exceed the limit values during the overhaul period, for this reason, asphalt concrete becomes an ineffective repair material.

2. Analysis of recent researches and problem statement

In many countries of the CIS and the former USSR, most roads were designed for load group B (60 kN per axle) and group A (100 kN per axle). However, at present, regulatory documents already provide for loads up to 115-130 kN per axle, which is actively used by cargo carriers. [3].

Cement-concrete pavements are the most promising in terms of increasing load-bearing capacity, performance and durability of roads. Under present day heavy traffic conditions, rigid pavements are more economical, even if the capital cost of their construction is significantly higher than asphalt concrete [4]. That is, with the growing of road sections with rigid pavements, the task of repairing these pavement types is becoming increasingly important.

It should be noted that the pavement works like a slab on elastic ground, so the pavement is designed for flexural strength. Therefore, use of high-strength concretes for repair cannot be a complete solution to the problem due to their high brittleness to dynamic loads. High performance concretes also cannot use their full potential due to low W/C ratio (no more than 0,3) and, as a consequence, increase in shrinkage deformation, shrinkage cracks may appear, especially in the early stages of hardening [5].

Concrete with dispersed reinforcement is widely used in road construction, which is used as different types of fiber: steel, basalt, polypropylene, glass [6,7]. The use of fiber makes it possible to increase abrasion resistance, crack resistance and the most important indicator for rigid pavements – flexural strength.

Steel-fiber concrete was first used as a repair material for rigid road pavements in the USA. [8]. Strength indicators of steel-fiber reinforced concrete in flexural strength reach high values > 12 MPa [9]. In turn, use of steel-fiber allows to reduce shrinkage deformations and increase frost resistance of concrete. The above-described advantages of this material make it possible to use it as a repair material for rigid road pavements.

To achieve the greatest structural and technological efficiency of repair compositions, to increase the early strength of concrete is crucial task. Its solution makes it possible to open traffic on the road section after repairs in a shorter time.

3. Research objective

Determination of the steel anchor fiber, hardening accelerator and superplasticizer amount on the compressive and flexural strength of modified repair concrete at the age of 2 and 28 days.

4. Materials and research methods

To increase workability, all concrete mixtures were modified with BASF MasterGlenium SKY 608 plasticizer in the amount of 1,2% by cement mass. This amount was considered rational based on the results of previous studies [10].

Cement CEM II/A-S 42,5 "Dyckerhoff Cement Ukraine" was used as a binding material. Crushed breakstone (grade 5-20 mm) and quartz sand were used as coarse and fine aggregates. The aggregates grain size is shown in table 1 and meets the requirements of the following standards: DSTU B V.2.7-75-98, DSTU B V.2.7-32-95, BS EN 12620:2013, ASTM C 33/C33M-18.

Table 1. Comparing of fine and coarse aggregates sieve analysis according to different building codes

Sieve size, mm	% Passing by weight		DSTU B V.2.7-75-98	DSTU B V.2.7-32-95	EN 12620:2013		ASTM C 33/C33M-18	
	Crushed breakstone	Sand	Crushed breakstone	Sand	Crushed breakstone	Sand	Crushed breakstone	Sand
20	100	100	≥ 90	-	90-99	-	90-100	-
10	54,7	100	20-70	-	-	-	20-55	100
5	5,0	100	0-10	-	0-15	100	0-10	95-100
2,5	0,8	97,0	-	80-100	0-5	85-99	0-5	80-100
1,25	0	84,5	-	55-95	-	-	-	50-85
0,63	0	32,6	-	30-80	-	-	-	25-60
0,315	0	35,3	-	20-50	-	-	-	5-30
0,16	0	10,5	-	0-15	-	-	-	0-10
< 0,16	0	1,5	-	-	-	-	-	-

According to the optimal symmetrical 9-point plan, 2-factors experiment [11] was carried out, the following compositional factors were varied:

X_1 – the amount of admixture Sika Rapid 3 hardening accelerator, from 0 to 2,4% by cement content (from 0 to 9,6 kg/m^3);

X_2 – the amount of steel anchor fiber with Ø 1 mm and length 50 mm, from 0 to 100 kg/m^3.

The concrete mixture was adjusted depending on the amount of dispersed reinforcement and modifiers. Mathematical planning of the experiment made it possible to minimize labor costs to determine the optimal concrete compositions and fiber-reinforced concrete mixtures with the best strength characteristics. Table 2 shows the matrix of a two-factor experiment with the investigated repair concrete compositions. The varied factor levels of experimental-statistical (ES) models were normalized in the range from -1 to +1.

Table 2. Concrete and fiber-concrete mixtures. Physical and mechanical properties of concretes

	Concrete and fibrous concrete mixtures											
№ of mixture	Cement, kg/m^3	Sand, kg/m^3	Crushed stone, (l/m^3)	SKY 608, %	Sika Rapid 3, % (X_1)	Steel fiber kg/m^3, (X_2)	$f_{ck.cube2}$ (MPa)	$f_{ck.cube28}$ (MPa)	f_{ctk2} (MPa)	f_{ctk28} (MPa)	W/C ratio	Average density, kg/m^3
1		830	1190		0	0	46,5	85,5	5,6	9,0	0,288	2530
2		830	1190		1,2	0	51,8	76,2	5,8	7,2	0,316	2540
3		830	1190		2,4	0	56,4	72,8	6,0	6,9	0,307	2530
4		788	1180		0	50	49,2	90,7	8,2	16,1	0,332	2530
5	400	786	1180	1,2	1,2	50	53,7	85,3	8,4	15,5	0,328	2550
6		782	1180		2,4	50	59,1	83,1	8,7	14,9	0,325	2540
7		752	1170		0	100	52,6	92,5	8,6	16,8	0,344	2550
8		750	1170		1,2	100	55,9	87,1	8,9	15,7	0,337	2590
9		745	1170		2,4	100	61,5	84,7	9,2	15,2	0,331	2550

5. Research results

For each concrete mixture, compressive strength test of samples at the age of 2 and 28 days were carried out (cubes 10x10x10 cm) and flexural strength test (prisms 10x10x40 cm) according to DSTU B V.2.7-214:2009.

Strength characteristics of mixture № 1 ($f_{ck.cube.2}$ = 46,5 MPa, $f_{ctk.2}$ = 5,6 MPa, $f_{ck.cube.28}$ = 46,5 MPa, $f_{ctk.28}$ = 9,0 MPa) made it possible to use it as a starting point for further improvement of concrete repair mixtures based on the research carried out [10,12]. Workability of the investigated concrete and steel-fiber concrete mixtures satisfied the requirements for full-depth repair compositions [13,14], slump was 6-9 cm.

Based on experimental data, ES-model was obtained (1), which describes the effect of the accelerator amount (x_1) and fiber (x_2) on the mixtures W/C ratio with equal workability. According to the model (1), a diagram was drawn in (figure 1) to analyze the combined effect of varied composition factors on the W/C ratio of repair concrete and steel-fiber concrete mixtures.

$$W/C = 0{,}328 - 0{,}005x_1 \pm 0x_1^2 \pm 0x_1x_2 + 0{,}012x_2 - 0{,}003x_2^2 \quad (1)$$

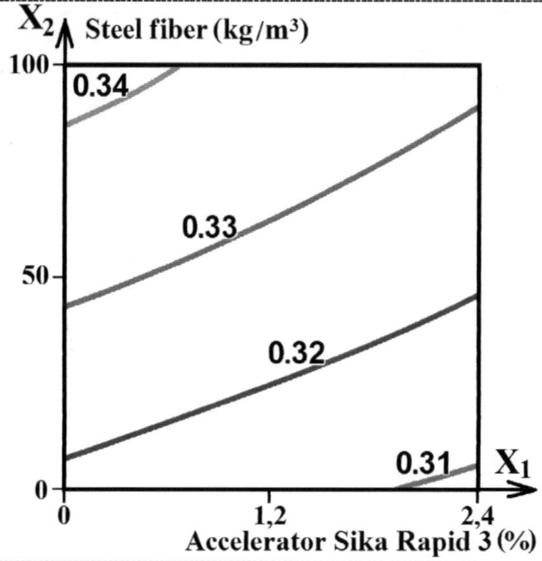

Figure 1. Influence of the hardening accelerator and steel fiber amount on the W/C ratio of concrete mixtures

Analysis of the diagram in figure 1 shows that dispersed reinforcement with steel fiber has a nonlinear effect on the water demand of the mixture, with fiber amount of 50 kg/m^3, W/C ratio increases more noticeably than with increase in the fiber dosage from 50 to 100 kg/m^3. However, even at the maximum dosage of fiber (100 kg/m^3), W/C ratio of mixtures are in the range of 0,332-0,344, which ensures quality of the repair composite. It is also necessary to note the effect of reducing the W/C ratio with an increase in the amount of the hardening accelerator agent. This is due to the accelerator is a liquid that can perform a function similar to water.

According to the data given in table 2, ES-models were drawn, reflecting the influence of various factors on the compressive strength at the age of 2 (2) and 28 (3) days of hardening.

$$f_{ck.cube.2}\,(\mathrm{MPa}) = 53{,}80 + 4{,}78x_1 + 0{,}42x_1^2 - 0{,}25x_1x_2 + 2{,}55x_2 \pm 0x_2^2 \quad (2)$$

$$f_{ck.cube28}\,(\mathrm{MPa}) = 85{,}0 - 4{,}68x_1 + 2{,}02x_1^2 + 1{,}23x_1x_2 + 4{,}97x_2 - 3{,}23x_2^2 \quad (3)$$

The combined effect of variable composition factors on the compressive strength of concrete and fiber-reinforced concrete is reflected in the diagrams drew using ES models (2), (3): at the age of 2 days (figure 2, a), at the age of 28 days (figure 2, b).

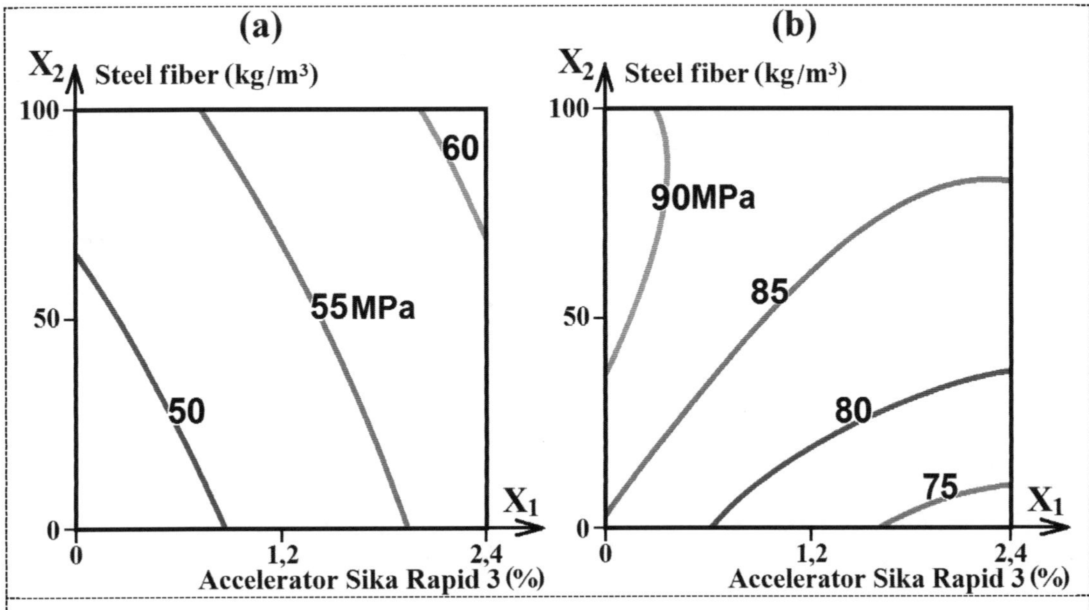

Figure 2. Influence of the hardening accelerator and steel fiber amount on the compressive strength at the age of 2 days (a) and 28 days (b)

Diagram on figure 2 (a) shows that, using the maximum amount of the hardening accelerator, regardless of the fiber amount, after 2 days of hardening, concrete has strength at least 55 MPa. That is, already at early age, the concrete strength actually corresponds to the C 32/40 grade. At the same time, fiber-reinforced concretes with the fiber amount in the composition of 50 kg/m^3 and higher have strength 55 MPa and more already with the amount of hardening accelerator from 1,4%. Using the maximum amount of dispersed reinforcement and Sika Rapid 3 admixture, early strength of fiber-reinforced concrete is at least 60 MPa, that is, already at the age of 2 days, strength is achieved, which provides concrete grade C 35/45.

Analysis of the diagram on figure 2 (b) shows that use of the hardening accelerator makes it possible to significantly increase the early strength of concretes for the road pavements repair, but negatively effects on the composite strength at the standard (design) age of 28 days [15]. But it can be stated that the compressive strength of all concretes and fiber-reinforced concretes investigated in the framework of this experiment at the design age was higher than required by standard documents for rigid road pavement concretes [16]. It is important to note that some reduction in compressive strength at 28 days with Sika Rapid 3 is fully compensated by the dispersed steel fiber reinforcement. Fibrous concrete with the 2,4% of hardening accelerator and steel fiber of 100 kg/m^3 has strength 85 MPa, which is similar to the design concrete strength without hardening accelerator and fiber.

The pavement works like a slab on elastic ground, so it is especially important to take into account the flexural strength for selecting repair mixtures. Based on the data given in table 2, ES-models were drawn, reflecting the influence of variable composition factors on flexural strength at the age of 2 (4) and 28 (5) days of hardening.

$$f_{ctk.2}\,(MPa) = 8,42 + 0,26x_1 + 0,02x_{12} + 0,05x_1x_2 + 1,55x_2 - 1,08x_{22} \quad (4)$$

$$f_{ctk28}\,(MPa) = 15,27 - 0,81x_1 + 0,35x_{12} \pm 0x_1x_2 + 4,10x_2 - 3,70x_{22} \quad (5)$$

On the basis of ES-models (4) and (5), diagrams were drawn showing the effect of the hardening accelerator admixture and steel fiber amount on the concrete flexural strength at early (figure 3 a) and standard (design) age (figure 3 b).

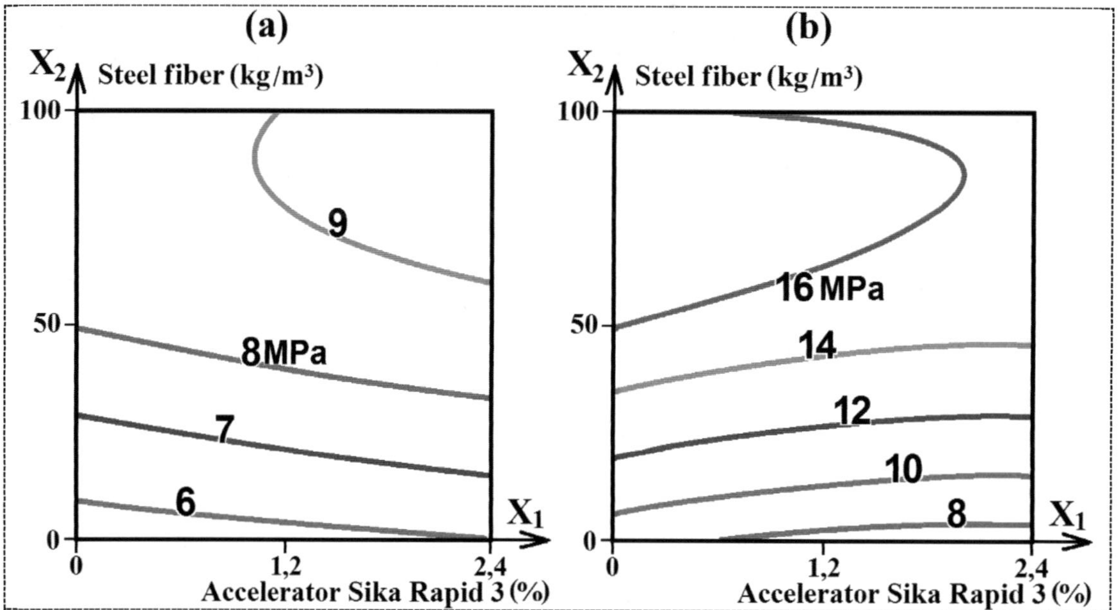

Figure 3. Influence of the hardening accelerator and steel fiber amount on the flexural strength of the studied concretes and fibrous concretes at the age of 2 days (a) and 28 days (b)

Analysis of the diagram (figure 3 a), shows that with the hardening accelerator Sika Rapid 3 in the fiber-reinforced concrete mixture from 1,2% and with steel fiber in the amount of 60 kg/m^3 and higher, early flexural strength of the composite is not less 8,5 MPa. The highest values (8,7..9,3 MPa) value $f_{ctk.2}$ reaches at fiber dosage in the range of 85..90 kg/m^3, further increase in the fibers dosage has small effect on the early flexural strength. Taking into account the compressive strength, which is described above, already after 2 days' completion of repair work using modified fiber-reinforced concrete, in most cases, it is possible to open the road traffic.

Figure 3 (b) shows that with the hardening accelerator agent is added to the concrete mixture, its flexural strength at the design age decreases similarly to the compressive strength. However, the most important fact is that the use of anchor steel fiber in amount of more than 60 kg/m^3, flexural strength of fiber-reinforced concretes not less than 16 MPa. Such high efficiency of the fiber at the design age is explained by the fact that the quality of the fibers work is largely due to their adhesion to the cement-sand matrix, and with age, the concrete and its matrix strength increases [17].

6. Conclusions and further researches

The study of the modified fiber-reinforced concrete strength (with the amount of MasterGlenium SKY 608 superplasticizer 1,2% by cement mass) showed that using a rational amount of the hardening accelerator Sika Rapid 3 (1,4-2,4% by cement mass) and steel anchor fiber (70..90 kg/m^3) the obtained composites are characterized by high compressive and flexural strength at early age and at the 28 days age. This allows to assert high efficiency of the developed modified fiber-reinforced concretes as a material for the rigid pavements repair, which allows starting the road pavement operation in the shortest possible time.

Further researches will be devoted to determining such important indicators as frost and wear resistance, which in turn effect on the concrete durability. It is also planned to conduct an experiment to determine and prevent shrinkage of high-early strength concretes and steel fiber reinforced concretes, especially in the early ages of hardening.

7. References

[1] Laszlo G, Zsolt B 2018 *Long-life pavements – European and American perspectives* (NBM&CW (New building materials & construction world), vol.24. Issue 3) p 122-135

[2] Official Site of the Highway Department of Ukraine - https://ukravtodor.gov.ua/

[3] Mozghovyi V, Besarab O, Kaskiv V, Mozghovyi O 2010 *Designing of overhaul of road surfaces* Highways and road construction **79** p 50-63

[4] Laurent G 2004 *Economic comparison between concrete and conventional road pavements in France* (9th International Symposium on Concrete Roads, Istanbul, Theme 1) p 21-31

[5] Usherov-Marshak A 2006 *Concrete admixtures: progress and challenges* Building materials **10** p 8-12

[6] Darter M 2017 *Concrete repair best practices: a series of case studies* (Applied research associates, Inc., Report № cmr17-013) p 119

[7] Harrington D, Fick G, Taylor P 2014 *Preservation and rehabilitation of urban concrete pavements using thin concrete overlays: solutions for joint deterioration in cold weather states* (National Concrete Pavement Technology Center, Iowa State Universit., Inc., Report № InTrans project 09-361) p 40

[8] Babych E, Andriichuk O, Uzhehov S, Shapoval I 2015 *The use of steel fiber reinforced concrete in road construction* Modern technologies and calculation methods in construction **4** p 3-9

[9] Dvorkin L, Babich E, Zhitkovskiy V, Bordyuzhenko O 2017 High-strength fast-hardening concretes and fiber-reinforced concretes Rivne NUVGP p 331

[10] Kryzhanovskiy V, Kroviakov S 2020 *Strength of rigid pavement concretes modified with polycarboxylate admixture on different types of cement* Bulletin of the Odessa State Academy of Civil Engineering and Architecture **79** p 92-98 (DOI: 10.31650/2415-377X-2020-79-92-98)

[11] Jeff Wu C F, Hamada M S 2009 *Experiments: Planning, Analysis, and Optimization* Wiley & Sons p 552

[12] Kryzhanovskyi V O, Kroviakov S O, Zavoloka M V 2021 *Influence of metakaolin on properties of concrete modified with polycarboxylate admixture for rigid pavement repair cement* Bulletin of the Odessa State Academy of Civil Engineering and Architecture p 90-97 (DOI: 10.31650/2415-377X-2021-82-90-97)

[13] Official Site of the Federal Highway Administration, Full-Depth Repairs - https://www.fhwa.dot.gov/pavement/concrete/full.cfm

[14] Raymond Bassim, Mohsen A Issa 2020 *Maturity-based estimates of concrete strength for Portland concrete cement pavements and patches at early age of opening to traffic* ACI materials journal p 197-211 (DOI: 10.14359 /51719081)

[15] Brikov A S 2008 Hydration of Portland cement SPb. SPbGTI (TU) p 30

[16] State building codes B.2.3-37641918-557: 2016 Highways. Road rigid pavement. Designing

[17] Marushchak U, Sanytsky M, Mazurak T, Olevych Y 2016 *Research of nanomodified Portland cement compositions with high early age strength* Eastern-European journal of enterprise technologies p 50-57 (DOI: 10.15587/1729-4061.2016.84175)

Development of an analytical method for calculating beams on a variable elastic Winkler foundation

Yu Krutii[1,2], M Surianinov[1], S Petrash[1], M Yezhov[1]

[1] Odessa State Academy of Civil Engineering and Architecture, 4, Didrikhsona st., Odessa, 65029, Ukraine
[2] Address all correspondence to this author. E-mail: yurii.krutii@gmail.com

Abstract. An analytical method has been developed for calculating beams lying on a continuous elastic Winkler foundation with a variable coefficient of subgrade resistance. The method is based on the exact solution of the differential equation of beam bending for the case when the bed coefficient is an arbitrary continuous function. In an analytical form, formulas for displacements and internal efforts are obtained. An analytical method for the numerical implementation of the exact solutions found is proposed.

1. Introduction

In engineering practice, beam elements of structures lying on a solid elastic foundation are often found. Such structures can include railway sleepers, strip foundations of buildings, foundations of dams, various pipelines laid on the ground and others.

Among the soil base models, the most common is the Winkler model, or the coefficient of subgrade resistance hypothesis. At the same time, there are a number of modifications of this model, which in an integral form make it possible to take into account the inhomogeneous properties of the foundation. The most common modification is the variable coefficient of the subgrade resistance model. In particular, such a model has found wide application in calculating the stress-strain state of foundations of structures lying on loess soils, which are characterized by subsidence. In this case, the coefficient of subgrade resistance is a variable, depending on the coordinate in which the settlement of the surface of the foundation is determined. From a mathematical point of view, this leads to the need to solve the corresponding differential equations with variable coefficients. A universal method for solving such equations is currently lacking. Perhaps this is why the calculation of beams on a variable elastic foundation is rarely found in the scientific literature.

Among the articles, which are devoted to the bending of a beam on an elastic Winkler base with a variable foundation coefficient, we note publications [1-5]. The article [1] is devoted to the case of a nonlinear, minus four power variation of the stiffness coefficient, allowing for closed-form representations. In the [2] the special case of a linear variation of the foundation coefficient is assumed. The analytical solution of the governing Ordinary Differential Equation is derived and represented in explicit closed-form in terms of generalized hypergeometric functions. The article [3] presents an analytical method, accompanied by a numerical scheme. The method employs a Green's function formulation, which results in a system of nonsingular integral equations. These equations can be discretized to yield a system of linear algebraic equations that can be solved by elementary numerical techniques. Authors [4] showed, that the solution may be achieved using only a small number of elements along the beam. A numerical example demonstrates the efficiency and accuracy of the procedure. In article [5] the case when the foundation coefficient is a power function of the coordinate is considered. Herein, an analytical solution is obtained and evaluated for the case when power is equal to 1. The numerical technique is developed for obtaining two of the solutions for other values of power.

As analysis shows, the development of an analytical method for calculating beams on a variable elastic Winkler base with a variable foundation coefficient is relevant.

2. Object of study, basic notation and the bending equation

The object of study is a beam of constant stiffness EI lying on a continuous variable elastic foundation, for which the Winkler hypothesis is accepted. Fig. 1 shows the design diagram of the beam, where $q(x)$ is a given distributed variable transverse load, $y(x)$ – deflection, $\varphi(x)$ – angle of rotation of the beam. Fig. 2 shows the internal forces arising in the cross sections of the beam element, namely, the bending moment $M(x)$ and the transverse force $Q(x)$.

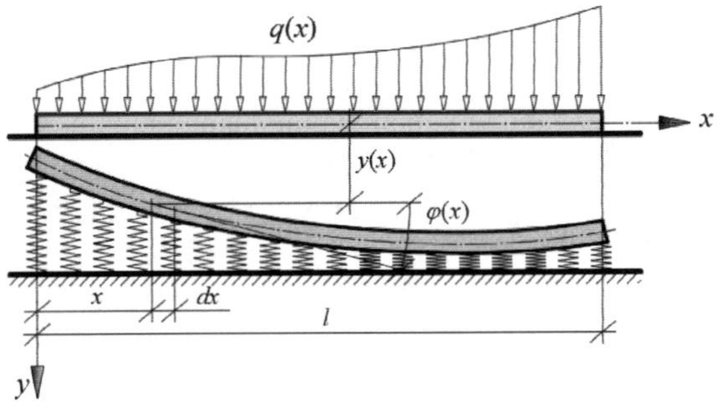

Figure 1. Beam design

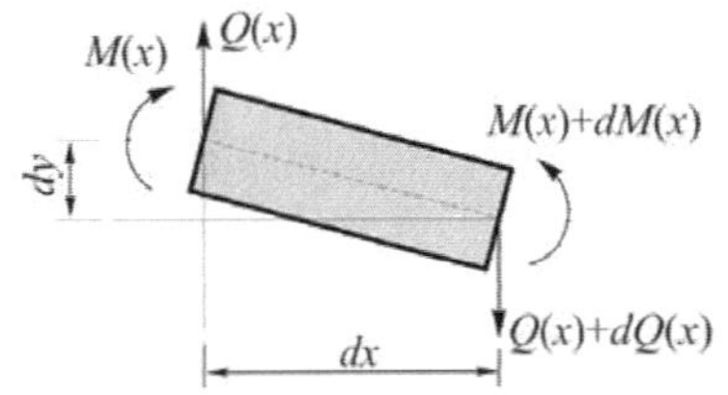

Figure 2. Internal forces

The stress-strain state of the beam is completely described by four components, namely, displacements $y(x)$, $\varphi(x)$ and internal forces $M(x), Q(x)$. Further we will call them beam state parameters.

According to Winkler's hypothesis, the reaction of the base to the beam $R(x)$ and the deflection of the beam $y(x)$ are interconnected by the equality $R(x) = -k(x)y(x)$, where $k(x)$ is a continuous variable linear coefficient of subgrade resistance. Relatively $k(x)$, we take the notation form $k(x) = k_0 B(x)$, where k_0 is the value of the linear coefficient of the subgrade resistance at some characteristic point of the beam (for example, at a point $x = 0$); $B(x)$ is dimensionless continuous function expressing the law of variation of the coefficient of subgrade resistance along the axis of the beam.

We will write a similar representation also for a given load: $q(x) = q_0 C(x)$, where q_0 is the load at some characteristic point of the beam; $C(x)$ is the dimensionless continuous function that sets the law of change of load from the coordinate x.

The differential equation of beam bending in our case takes the form

$$EI\, y''''(x) + k_0 B(x) y(x) = q_0 C(x)\,. \quad (1)$$

After the deflection function $y(x)$ is found from this equation, the remaining state parameters are found from the differential dependences known from the theory of beam bending:

$$\varphi(x) = y'(x);\ M(x) = -EI y''(x);\ Q(x) = -EI y'''(x)\,.$$

The aim of the work is to develop an analytical method for calculating the beam bending under the conditions described above. To achieve the goal, the following tasks are formulated and solved:

- find the exact solution to the equation (1);

- get formulas for beam state parameters;
- indicate the analytical method for the numerical implementation of the exact solutions found.

3. Materials and Methods

Earlier in [6], the method of direct integration of differential equations with continuous variable coefficients was developed. Using this method, exact solutions for a number of differential equations of statics and dynamics were also constructed there. The essence of the proposed method can also be considered, for example, by publications [7-10], where a number of problems in the mechanics of a deformable solid are solved with its help.

In this work, the exact solution of the differential bending equation (1), which formed the basis of the study, was constructed by direct integration.

4. Exact solution to the bending equation and formulas for beam state parameters

We denote by the $X_n(x)$ $(n=1,2,3,4)$ sought dimensionless fundamental solutions of the homogeneous equation

$$EI\, y''''(x)+k_0 B(x)y(x)=0\,, \tag{2}$$

and through

$$X^*(x)=\frac{q_0 l^4}{EI}X_5(x) \tag{3}$$

– partial solution of an inhomogeneous equation (1), where $X_5(x)$ – sought dimensionless function.

Along with the bending equation (1), we will also consider a system of differential equations equivalent to it. As a vector of system's unknown, we take a vector whose components are the beam state parameters

$$\Phi(x)=\begin{pmatrix} y(x) \\ \varphi(x) \\ M(x) \\ Q(x) \end{pmatrix}=\begin{pmatrix} y(x) \\ y'(x) \\ -EI\, y''(x) \\ -EI\, y'''(x) \end{pmatrix}. \tag{4}$$

In this case, the system will be written as

$$\frac{d\Phi(x)}{dx}=H(x)\Phi(x)-f(x)\,, \tag{5}$$

where

$$H(x)=\begin{pmatrix} 0 & 1 & 0 & 0 \\ 0 & 0 & -1/EI & 0 \\ 0 & 0 & 0 & 1 \\ k_0B(x) & 0 & 0 & 0 \end{pmatrix},\quad f(x)=\begin{pmatrix} 0 \\ 0 \\ 0 \\ q_0C(x) \end{pmatrix}. \tag{6}$$

The solutions of the homogeneous equation $X_n(x)\ (n=1,2,3,4)$ and the function $X_5(x)$ will be sought in the form of series in powers of the dimensionless parameter $-K$ with variable coefficients

$$X_n(x)=\beta_{n,0}(x)-K\beta_{n,1}(x)+K^2\beta_{n,2}(x)-K^3\beta_{n,3}(x)+\dots\ (n=1,2,3,4,5)\,, \tag{7}$$

where $K = k_0 l^4 / EI$. Moreover, following the terminology adopted in [6], the functions $\beta_{n,0}(x)$ will be called initial, and the functions $\beta_{n,k}(x)$ $(k = 1,2,3,...)$ – generators.

For now, we assume that series (7), as well as similar series composed of the first four derivatives of functions $\beta_{n,0}(x)$, $\beta_{n,k}(x)$ $(k = 1,2,3,...)$, converge uniformly. In this case, the operation of differentiating the rows will be possible.

The initial and generating functions will be sought from the conditions that the functions $X_n(x)$ $(n = 1,2,3,4)$ satisfy the homogeneous equation (2), and the function (3) – the inhomogeneous equation (1), that is, from the conditions that the equalities are satisfied:

$$X_n''''(x) + \frac{K}{l^4} B(x) X_n(x) = 0 \;\; (n = 1,2,3,4); \tag{8}$$

$$X_5''''(x) + \frac{K}{l^4} B(x) X_5(x) = \frac{1}{l^4} C(x). \tag{9}$$

Substituting into equalities (8), (9) instead of $X_n(x)$ their value (7), after transformations we get:

$$\beta_{n,0}''''(x) + \sum_{k=1}^{\infty} (-K)^k \left(\beta_{n,k}''''(x) - \frac{1}{l^4} B(x) \beta_{n,k-1}(x) \right) = 0 \;\; (n = 1,2,3,4); \tag{10}$$

$$\beta_{5,0}''''(x) + \sum_{k=1}^{\infty} (-K)^k \left(\beta_{5,k}''''(x) - \frac{1}{l^4} B(x) \beta_{5,k-1}(x) \right) = \frac{1}{l^4} C(x). \tag{11}$$

Hence, equating to zero all the coefficients for the degrees of the parameter $-K$, including the zero degree, we arrive at the following equations with respect to the initial and generating functions:

$$\beta_{n,0}''''(x) = 0 \;\; (n = 1,2,3,4); \tag{12}$$

$$\beta_{5,0}''''(x) = \frac{1}{l^4} C(x); \tag{13}$$

$$\beta_{n,k}''''(x) = \frac{1}{l^4} B(x) \beta_{n,k-1}(x) \;\; (n = 1,2,3,4,5). \tag{14}$$

Before integrating these equations, we define the boundary conditions:

$$\begin{pmatrix} \beta_{1,0}(0) & \beta_{2,0}(0) & \beta_{3,0}(0) & \beta_{4,0}(0) \\ \beta_{1,0}'(0) & \beta_{2,0}'(0) & \beta_{3,0}'(0) & \beta_{4,0}'(0) \\ \beta_{1,0}''(0) & \beta_{2,0}''(0) & \beta_{3,0}''(0) & \beta_{4,0}''(0) \\ \beta_{1,0}'''(0) & \beta_{2,0}'''(0) & \beta_{3,0}'''(0) & \beta_{4,0}'''(0) \end{pmatrix} = \begin{pmatrix} 1 & 0 & 0 & 0 \\ 0 & 1/l & 0 & 0 \\ 0 & 0 & 1/l^2 & 0 \\ 0 & 0 & 0 & 1/l^3 \end{pmatrix}; \tag{15}$$

$$\beta_{5,0}(0) = \beta_{5,0}'(0) = \beta_{5,0}''(0) = \beta_{5,0}'''(0) = 0; \tag{16}$$

$$\beta_{n,k}(0) = \beta_{n,k}'(0) = \beta_{n,k}''(0) = \beta_{n,k}'''(0) = 0 \;\; (n = 1,2,3,4,5) \;\; (k = 1,2,3,...). \tag{17}$$

Note that conditions (15)-(17) are chosen in such a way as to ensure that the integration constants are equal zero.

As a result, after integration (12) - (14), we will have the following set of formulas:

$$\beta_{n,0}(x) = \frac{1}{(n-1)!} \left(\frac{x}{l} \right)^{n-1} \;\; (n = 1,2,3,4); \tag{18}$$

$$\beta_{5,0}(x) = \frac{1}{l^4} \int_0^x \int_0^x \int_0^x \int_0^x C(x) dx dx dx dx; \tag{19}$$

$$\beta_{n,k}(x)=\frac{1}{l^4}\int_0^x\int_0^x\int_0^x\int_0^x B(x)\beta_{n,k-1}(x)\,dxdxdxdx \;\; (n=1,2,3,4,5)(k=1,2,3,...)\,. \tag{20}$$

Thus, by means of recurrence formula (20), for each of the initial functions (18), (19), the corresponding generating functions are found. Obviously, for such functions, equalities (8), (9) are satisfied identically.

In addition to recurrence formula (20), the generating functions can be represented in expanded form

$$\beta_{n,k}(x)=\frac{1}{l^{4k}}\int_0^x\int_0^x\int_0^x\int_0^x B(x)....\int_0^x\int_0^x\int_0^x\int_0^x B(x)\int_0^x\int_0^x\int_0^x\int_0^x B(x)\,\beta_{n,0}(x)\,dxdxdxdx...dxdx \;\; (n=1,2,3,4,5)\,. \tag{21}$$

The right side of the last formula contains $4k$ integrals, not taking into account the integrals through which the initial function is expressed (19).

After the initial and generating functions are defined, it is necessary to prove the convergence of the series (7), as well as the series of derivatives. We give a proof here, for example, for the first four of the series (7). Assuming

$$g=\max_{x\in[0,l]} B(x)$$

and using any of formulas (20), (21), we obtain the following estimates for generating functions

$$\beta_{n,k}(x)\le\frac{g^k}{(n+4k-1)!}\left(\frac{x}{l}\right)^{n+4k-1} \;\; (n=1,2,3,4)\;(k=1,2,3,...)\,.$$

Using these estimates, we obtain

$$\left|X_n(x)\right|\le\sum_{k=0}^{\infty}K^k\beta_{n,k}(x)\le\frac{1}{(\sqrt[4]{gK})^{n-1}}\sum_{k=0}^{\infty}\frac{1}{(n+4k-1)!}\left(\sqrt[4]{gK}\,\frac{x}{l}\right)^{n+4k-1}=$$

$$=\frac{1}{(\sqrt[4]{gK})^{n-1}}Y_n\left(\sqrt[4]{gK}\,\frac{x}{l}\right) \;\; (n=1,2,3,4)\,, \tag{22}$$

where $Y_n\left(\sqrt[4]{gK}\,\frac{x}{l}\right)$ – A.N. Krylov functions, which are defined by the equalities:

$$Y_1(z)=\frac{1}{2}(\cosh z+\cos z);\; Y_2(z)=\frac{1}{2}(\sinh z+\sin z)\,;$$

$$Y_3(z)=\frac{1}{2}(\cosh z-\cos z);\; Y_2(z)=\frac{1}{2}(\sinh z-\sin z)\,.$$

Based on (22), we conclude that the series (7) for the values $n=1,2,3,4$ converge absolutely and uniformly. Similarly, the convergence of the remaining series is proved.

Thus, formulas (7), (18)-(21) define four solutions of the homogeneous equation (2) and a particular solution of equation (1).

Each of the solutions $X_n(x)\;(n=1,2,3,4)$ generates a vector by formula (4)

$$\Phi_n(x)=\begin{pmatrix} X_n(x) \\ X'_n(x) \\ -EI\,X''_n(x) \\ -EI\,X'''_n(x) \end{pmatrix} \;\; (n=1,2,3,4)$$

– solving a homogeneous system of differential equations corresponding to (5). Then the matrix

$$\Omega(x)=\begin{pmatrix} X_1(x) & X_2(x) & X_3(x) & X_4(x) \\ X_1'(x) & X_2'(x) & X_3'(x) & X_4'(x) \\ -EI\,X_1''(x) & -EI\,X_2''(x) & -EI\,X_3''(x) & -EI\,X_4''(x) \\ -EI\,X_1'''(x) & -EI\,X_2'''(x) & -EI\,X_3'''(x) & -EI\,X_4'''(x) \end{pmatrix}, \quad (23)$$

composed of these vectors will also satisfy a homogeneous system. Moreover, taking into account (15), (17), we find:

$$\Omega(0)=\begin{pmatrix} 1 & 0 & 0 & 0 \\ 0 & 1/l & 0 & 0 \\ 0 & 0 & -EI/l^2 & 0 \\ 0 & 0 & 0 & -EI/l^3 \end{pmatrix}; \quad |\Omega(0)|=\frac{(EI)^2}{l^6}\neq 0 .$$

Therefore, matrix (23) is the fundamental matrix [11] for system (5), and $X_n(x)$ $(n=1,2,3,4)$ is the fundamental solutions of equation (2).

Multiplying on the right $\Omega(x)$ by a constant matrix $\Omega^{-1}(0)$, we obtain a new fundamental matrix $\Lambda(x)=\Omega(x)\Omega^{-1}(0)$ for which

$$\Lambda(0)=\begin{pmatrix} 1 & 0 & 0 & 0 \\ 0 & 1 & 0 & 0 \\ 0 & 0 & 1 & 0 \\ 0 & 0 & 0 & 1 \end{pmatrix}. \quad (24)$$

As is known [11], a fundamental matrix possessing property (24) is uniquely determined and is called a matrix. Moreover, the general solution of the homogeneous system corresponding to (5) is written in the form

$$\Phi(x)=\Lambda(x)\Phi(0) .$$

Forming with the help of (3) the vector

$$\Phi^*(x)=\frac{q_0 l^4}{EI}\begin{pmatrix} X_5(x) \\ X_5'(x) \\ -EI\,X_5''(x) \\ -EI\,X_5'''(x) \end{pmatrix},$$

we obtain a particular solution to the inhomogeneous system (5), which is easy to verify by substitution. Considering (16), (17), for this vector we also find

$$\Phi^*(0)=\begin{pmatrix} 0 \\ 0 \\ 0 \\ 0 \end{pmatrix}.$$

Therefore, the general solution of the inhomogeneous system of differential equations (5) can be written as follows

$$\Phi(x)=\Lambda(x)\Phi(0)+\Phi^*(x) . \quad (25)$$

IOP Conf. Series: Materials Science and Engineering 1162 (2021) 012009 doi:10.1088/1757-899X/1162/1/012009

In the calculations it is convenient to operate with dimensionless quantities. Based on the analysis of formulas (18)-(21) and the results of [6], we conclude that the initial and generating functions $\beta_{n,k}(x)$ $(n=1,2,3,4,5)$ $(k=0,1,2,...)$ are dimensionless. As a result, solutions $X_n(x)$ $(n=1,2,3,4,5)$ will also be dimensionless.

Before expanding elementwise equality (25), we introduce the functions:

$$\tilde{X}_n(x)=lX_n'(x);\ \bar{X}_n(x)=l^2X_n''(x);\ \hat{X}_n(x)=l^3X_n'''(x). \tag{26}$$

It follows directly from formulas (7), (26) that these functions are represented by series:

$$\tilde{X}_n(x)=\tilde{\beta}_{n,0}(x)-K\tilde{\beta}_{n,1}(x)+K^2\tilde{\beta}_{n,2}(x)-K^3\tilde{\beta}_{n,3}(x)+...\ (n=1,2,3,4,5); \tag{27}$$

$$\bar{X}_n(x)=\bar{\beta}_{n,0}(x)-K\bar{\beta}_{n,1}(x)+K^2\bar{\beta}_{n,2}(x)-K^3\bar{\beta}_{n,3}(x)+...\ (n=1,2,3,4,5); \tag{28}$$

$$\hat{X}_n(x)=\hat{\beta}_{n,0}(x)-K\hat{\beta}_{n,1}(x)+K^2\hat{\beta}_{n,2}(x)-K^3\hat{\beta}_{n,3}(x)+...\ (n=1,2,3,4,5), \tag{29}$$

where

$$\tilde{\beta}_{n,0}(x)=l\beta_{n,0}'(x),\ \bar{\beta}_{n,0}(x)=l^2\beta_{n,0}''(x),\ \hat{\beta}_{n,0}(x)=l^3\beta_{n,0}'''(x)\ (n=1,2,3,4,5), \tag{30}$$

$$\tilde{\beta}_{n,k}(x)=l\beta_{n,k}'(x),\ \bar{\beta}_{n,k}(x)=l^2\beta_{n,k}''(x),\ \hat{\beta}_{n,k}(x)=l^3\beta_{n,k}'''(x)\ (n=1,2,3,4,5)\ (k=0,1,2,...). \tag{31}$$

The functions (30) here act as initial ones, and the functions (31) as the generating functions for the corresponding series (27)-(29).

Unlike derivative functions $X_n(x)$, functions (26) will be dimensionless [6].

Now revealing equality (25) and taking into account (26), we obtain the final formulas for the beam state parameters expressed in terms of the initial parameters and dimensionless functions:

$$y(x)=y(0)X_1(x)+\varphi(0)lX_2(x)-M(0)\frac{l^2}{EI}X_3(x)-Q(0)\frac{l^3}{EI}X_4(x)+\frac{q_0l^4}{EI}X_5(x); \tag{32}$$

$$\varphi(x)=y(0)\frac{1}{l}\tilde{X}_1(x)+\varphi(0)\tilde{X}_2(x)-M(0)\frac{l}{EI}\tilde{X}_3(x)-Q(0)\frac{l^2}{EI}\tilde{X}_4(x)+\frac{q_0l^3}{EI}\tilde{X}_5(x); \tag{33}$$

$$M(x)=-y(0)\frac{EI}{l^2}\bar{X}_1(x)-\varphi(0)\frac{EI}{l}\bar{X}_2(x)+M(0)\bar{X}_3(x)+Q(0)l\bar{X}_4(x)-q_0l^2\bar{X}_5(x); \tag{34}$$

$$Q(x)=-y(0)\frac{EI}{l^3}\hat{X}_1(x)-\varphi(0)\frac{EI}{l^2}\hat{X}_2(x)+M(0)\frac{1}{l}\hat{X}_3(x)+Q(0)\hat{X}_4(x)-q_0l\hat{X}_5(x). \tag{35}$$

Theoretically, formulas (32)-(35) allow one to rely on the bending of a beam on a variable elastic foundation under any boundary conditions. However, when implementing these formulas in practice, difficulties may arise associated with the need to calculate integral expressions (20) or (21), which determine the generating functions. Therefore, for the final solution of the problem, we represent the generating functions by power series.

5. Representation of generating functions by power series

One of the classical methods of integration is based on replacing the integrand by the Maclaurin series. In our case, this approach allows us to calculate the integral (21) in an analytical form.

So, we will proceed from the fact that dimensionless functions $B(x)$ and $C(x)$ are represented by power series:

$$B(x)=B_0+B_1\left(\frac{x}{l}\right)+B_2\left(\frac{x}{l}\right)^2+...+B_j\left(\frac{x}{l}\right)^j+...; \tag{36}$$

IOP Conf. Series: Materials Science and Engineering 1162 (2021) 012009 doi:10.1088/1757-899X/1162/1/012009

$$C(x) = C_0 + C_1\left(\frac{x}{l}\right) + C_2\left(\frac{x}{l}\right)^2 + \ldots + C_j\left(\frac{x}{l}\right)^j + \ldots, \tag{37}$$

where

$$B_0 = B(0),\ B_j = \frac{l^j B^{(j)}(0)}{j!},\ C_0 = C(0),\ C_j = \frac{l^j C^{(j)}(0)}{j!}\ (j = 1,2,3,\ldots)$$

– dimensionless coefficients, and the index (j) means the order of the derivative. Then, taking into account (37), the initial function (19) after integration will appear in the form of a series

$$\beta_{5,0}(x) = \left(\frac{x}{l}\right)^4 \sum_{j=0}^{\infty} \frac{C_j}{(j+1)(j+2)(j+3)(j+4)}\left(\frac{x}{l}\right)^j. \tag{38}$$

Directly from formula (20) or (21), taking into account (18), (36)-(38), we conclude that the generating functions $\beta_{n,k}(x)\ (n = 1,2,3,4,5)(k = 1,2,3,\ldots)$ will also be power series. Moreover, since formula (21) contains $4k$ integrals, the degree of the first term of the indicated series will be equal to $n + 4k - 1$.

Therefore, the generating functions can be represented as

$$\beta_{n,k}(x) = \left(\frac{x}{l}\right)^{n+4k-1} \sum_{j=0}^{\infty} c_{n,k,j}\left(\frac{x}{l}\right)^j\ (n = 1,2,3,4,5)(k = 1,2,3,\ldots), \tag{39}$$

where $c_{n,k,j}$ – coefficients to be determined. Then

$$\beta_{n,k-1}(x) = \left(\frac{x}{l}\right)^{n+4k-5} \sum_{j=0}^{\infty} c_{n,k-1,j}\left(\frac{x}{l}\right)^j\ (n = 1,2,3,4,5)(k = 1,2,3,\ldots). \tag{40}$$

Moreover, for the product of series (36) and (40) we obtain

$$B(x)\beta_{n,k-1}(x) = \left(\frac{x}{l}\right)^{n+4k-5} \sum_{j=0}^{\infty} d_{n,k-1,j}\left(\frac{x}{l}\right)^j, \tag{41}$$

where

$$d_{n,k-1,j} = \sum_{i=0}^{j} B_{j-i}\, c_{n,k-1,i}. \tag{42}$$

We substitute in the formula (20) instead of $\beta_{n,k}(x)$ and $B(x)\beta_{n,k-1}(x)$ their values (39) and (41), after which we perform the integration. As a result, we arrive at equality

$$\sum_{j=0}^{\infty} c_{n,k,j}\left(\frac{x}{l}\right)^j = \sum_{j=0}^{\infty} \frac{d_{n,k-1,j}}{p_{n,k,j}}\left(\frac{x}{l}\right)^j\ (n = 1,2,3,4,5)(k = 1,2,3,\ldots),$$

where

$$p_{n,k,j} = (n + 4k + j - 4)(n + 4k + j - 3)(n + 4k + j - 2)(n + 4k + j - 1).$$

From this, taking into account (42), we find

$$c_{n,k,j} = \frac{1}{p_{n,k,j}} \sum_{i=0}^{j} B_{j-i}\, c_{n,k-1,i}\ (n = 1,2,3,4,5)(k = 1,2,3,\ldots)(j = 0,1,2,\ldots). \tag{43}$$

The formula found for the desired coefficients is recursive in index k. For complete certainty, such a formula needs to be supplemented with initial values $c_{n,0,j}\,(n=1,2,3,4,5)(j=0,1,2,...)$. For this purpose, we put $k=1$ in formula (40) and compare the result in turn with formulas (18) and (38). As a result, we will have:

$$c_{n,0,0}=\frac{1}{(n-1)!};\; c_{n,0,j}=0\;\; (n=1,2,3,4)(j=1,2,3,...)\,; \tag{44}$$

$$c_{5,0,j}=\frac{C_j}{p_{5,0,j}}\;\;(j=0,1,2,...)\,. \tag{45}$$

Thus, the dimensionless coefficients of series (39) are completely determined by formulas (43)-(45). As a result, the generating functions $\beta_{n,k}(x)\;(n=1,2,3,4,5)(k=1,2,3,...)$ are represented by power series.

Finally, we give a list of formulas necessary for the numerical implementation of the method. Table 1 shows the initial functions for the values $n=1,2,3,4$.

Table 1. Initial functions

$\beta_{1,0}(x)=1$;	$\beta_{2,0}(x)=\frac{x}{l}$;	$\beta_{3,0}(x)=\frac{1}{2!}\left(\frac{x}{l}\right)^2$;	$\beta_{4,0}(x)=\frac{1}{3!}\left(\frac{x}{l}\right)^3$;
$\tilde{\beta}_{1,0}(x)=0$;	$\tilde{\beta}_{2,0}(x)=1$;	$\tilde{\beta}_{3,0}(x)=\frac{x}{l}$;	$\tilde{\beta}_{4,0}(x)=\frac{1}{2!}\left(\frac{x}{l}\right)^2$;
$\widehat{\beta}_{1,0}(x)=0$;	$\widehat{\beta}_{2,0}(x)=0$;	$\widehat{\beta}_{3,0}(x)=1$;	$\widehat{\beta}_{4,0}(x)=\frac{x}{l}$;
$\hat{\beta}_{1,0}(x)=0$;	$\hat{\beta}_{2,0}(x)=0$;	$\hat{\beta}_{3,0}(x)=0$;	$\hat{\beta}_{4,0}(x)=1$.

When $n=5$ the initial functions are determined by equalities (38), (46) - (48):

$$\tilde{\beta}_{5,0}(x)=\left(\frac{x}{l}\right)^3\sum_{j=0}^{\infty}\frac{C_j}{(j+1)(j+2)(j+3)}\left(\frac{x}{l}\right)^j; \tag{46}$$

$$\widehat{\beta}_{5,0}(x)=\left(\frac{x}{l}\right)^2\sum_{j=0}^{\infty}\frac{C_j}{(j+1)(j+2)}\left(\frac{x}{l}\right)^j; \tag{47}$$

$$\hat{\beta}_{5,0}(x)=\frac{x}{l}\sum_{j=0}^{\infty}\frac{C_j}{j+1}\left(\frac{x}{l}\right)^j. \tag{48}$$

For the generating functions, formulas (39), (49) - (51) hold:

$$\tilde{\beta}_{n,k}(x)=\left(\frac{x}{l}\right)^{n+4k-2}\sum_{j=0}^{\infty}\tilde{c}_{n,k,j}\left(\frac{x}{l}\right)^j\;\;(n=1,2,3,4,5)(k=1,2,3,...)\,; \tag{49}$$

$$\widehat{\beta}_{n,k}(x)=\left(\frac{x}{l}\right)^{n+4k-3}\sum_{j=0}^{\infty}\widehat{c}_{n,k,j}\left(\frac{x}{l}\right)^j\;\;(n=1,2,3,4,5)(k=1,2,3,...)\,; \tag{50}$$

$$\hat{\beta}_{n,k}(x)=\left(\frac{x}{l}\right)^{n+4k-4}\sum_{j=0}^{\infty}\hat{c}_{n,k,j}\left(\frac{x}{l}\right)^j\;\;(n=1,2,3,4,5)(k=1,2,3,...)\,, \tag{51}$$

where

$$\tilde{c}_{n,k,j}=c_{n,k,j}(n+4k+j-1)\,,$$

$$\hat{c}_{n,k,j} = c_{n,k,j}(n+4k+j-1)(n+4k+j-2),$$
$$\hat{c}_{n,k,j} = c_{n,k,j}(n+4k+j-1)(n+4k+j-2)(n+4k+j-3).$$

Thus, in essence, an analytical method is proposed for the numerical implementation of the exact solutions found.

6. Results and discussion

Whenever the study of a physical phenomenon is reduced to a differential equation, the key is the question of constructing its exact analytical solution. However, researchers along the way often encounter a well-known mathematical problem, which is the lack of a universal method for integrating differential equations with variable coefficients. As a rule, solutions can be obtained only for some special cases. Probably, this can explain the predominant use of approximate methods. The foregoing fully applies to the problem of bending the beam on a variable elastic foundation.

In this paper, for the case when the elastic base is described by the Winkler model with an arbitrary continuous variable coefficient of the subgrade resistance, these difficulties were overcome. As a result, the following results were obtained:

1. The exact solution of the differential equation of the bending of the beam is constructed.
2. Formulas are obtained for displacements and internal forces in the beam.
3. An analytical method for the numerical implementation of the exact solutions found is proposed.

In fact, the solution of the original problem is reduced only to the implementation of the given boundary conditions, the determination of unknown initial parameters and the numerical implementation of the solutions found.

This paper is theoretical in nature. The authors plan to devote the following publications to the application of the developed method to the calculation of real structures.

References

[1] Froio D, Rizzi E 2016 Analytical solution for the elastic bending of beams lying on a variable Winkler support *Acta Mechanica* **227.4** pp 1157-1179. doi.org/10.1007/s00707-015-1508-y

[2] Froio D, Rizzi E 2017 Analytical solution for the elastic bending of beams lying on a linearly variable Winkler support *International Journal of Mechanical Sciences* **128** pp 680-694 doi.org/10.1016/j.ijmecsci.2017.04.021

[3] Guo Y J, Weitsman Y J 2002 Solution method for beams on nonuniform elastic foundations *Journal of engineering mechanics* **128.5** pp 592-594 doi.org/10.1061/(ASCE)0733-9399(2002)128:5(592)

[4] Clastornik J, Eisenberger M, Yankelevsky D, Adin M A 1986 Beams on Variable Winkler Elastic Foundation *Journal of Applied Mechanics* **53.4** pp 925–928 doi.org/10.1115/1.3171882

[5] Franklin J N, Scott R F 1979 Beam equation with variable foundation coefficient *Journal of the Engineering Mechanics division* **105.5** pp 811-827

[6] Krutii Yu S 2016 *Development of a method for solving problems of stability and vibrations of deformable systems with variable continuous parameters* Sc.D. diss., Lutsk National Technical University

[7] Krutii Yu S 2018 Construction of a solution of the problem of stability of a bar with arbitrary continuous parameters *Journal of Mathematical Sciences* **231** pp 665-677 doi.org/10.1007/s10958-018-3843-8

[8] Shvab'yuk V I, Krutii Yu S, Sur'yaninov M G 2016 Investigation of the Free Vibrations of Bar Elements with Variable Parameters Using the Direct Integration Method *Strength of Materials* **48** pp 384-393 doi: 10.1007/s11223-016-9776-x

[9] Krutii Yu S, Sur'yaninov M G, Vandynskyi V 2018 Development of the method for calculation of cantilever construction's oscillations taking into account own weight *Eastern-European Journal of Enterprise Technologies* **3.7** pp 13-19 doi: 10.15587/1729-4061.2018.131165

[10] Krutii Yu, Sur'yaninov M, Vandynskyi V 2019 Analytic Formulas for the Cantilever Structures' Natural Frequencies with Taking into Account the Dead Weight *Materials Science Forum* **968** pp 450–459 doi.org/10.4028/www.scientific.net/MSF.968.450
[11] Gantmacher F R 1959 *The Theory of Matrices* Chelsea New York American Mathematical Soc

Change in effective viscosity and level of synergism of cement-containing mechanoactivated suspension

L Ksonshkevych[1,2], **I Barabash**[1,2], **K Streltsov**[1,2] **and O Krantovska**[1,3]
[1] Odessa State Academy of Civil Engineering and Architecture, 65029, Didrikhsona st., 4, Odessa, Ukraine
[2]Department of Urban Construction and Management
[3]Department of Materials Strength

E-mail: ksonshkevych@ogasa.org.ua

Abstract. The article considers the influence of prescription factors (ground chamot, superplasticizer C-3) on the change of the effective viscosity of mechanically activated cement-containing suspensions used in the technology of lightweight concrete mixtures. The synergistic effect of reducing the viscosity of cement-containing suspensions during their activation in high-speed mixers of turbulent type in the presence of C-3 was revealed. Optimized modes of mechanical activation of suspensions that provide the maximum reduction of their effective viscosity.

1. Introduction

Lightweight concrete mixtures are increasingly used in the concreting of reinforced concrete structures, especially in frame-monolithic construction [1]. Prerequisites for obtaining lightweight concrete mixtures are the use of polyfraction fillers and additives, such as chamot. The need to obtain lightweight concrete mixtures based on Portland cement and ordinary granite rubble makes us look for new technological techniques. In particular, partial replacement of granite rubble with expanded clay gravel and the use of separate technology for preparing concrete mixes using high-speed mixers. This technology will allow to obtain high-tech cement-containing suspensions with low viscosity, which provide increased easy-to-place mixtures with lower consumption of water [2 - 6].

2. Analysis of recent research and publications

Works [7, 8] found that the use of chamot as an additive to Portland cement, makes it possible to regulate both the easy-to-placing mixture and the processes of structure formation of cement stone and concrete based on it [8]. Strengthening the role of mineral and organic components is possible due to their mechanical activation in the suspension in high-speed mixers of the original design [2, 4]. The addition of surfactants to the suspension enhances the role of chamot in the formation of the structure of cement stone [9].

3. The purpose of the research

The aim is to investigate the influence of mechanoactivation modes, chamot concentration, C-3 superplasticizer on the change of effective viscosity of binder suspensions. The task of research is to obtain a synergistic effect of reducing the viscosity of cement-containing suspensions with the addition of chamot with the joint action of high-speed mixing, superplasticizer C-3.

4. Research methods

It was of interest to elucidate the effect of high-speed mixing on the change in the effective viscosity of a cement-containing suspension. Ground chamot was used as a mineral additive to Portland cement (the content of chamot in the binder varied from 0 to 30% of the consumption of Portland cement). Ground chamot was milled to three type of specific surface - 250, 350 and 450 m^2 / kg. Superplasticizer C-3 was used as a reducing additive, the concentration of which varied from 0 to 1% of the binder mass. The effective viscosity of cement-containing suspensions was determined using a rotary viscometer with coaxial cylinders. The initial (control) composition of the suspension was accepted as follows: Portland cement - 500g; water - 145g. The effective viscosity of such a suspension was equal to 2950 cP.

Given in table. 1 experimental data on effective viscosity indicate that mechanical activation of a suspension leads to a sharp decrease in its effective viscosity. This is especially evident for suspensions, in the composition of which the superplasticizer C-3 was introduced in an amount of 1% by weight of the binder.

Table 1. Influence of the activation time of Portland cement on the effective viscosity of suspensions

Portland cement, %	Chamot, %	Superplasticizer C-3, %	Specific surface area , , m^2/kg	The activation time, s						
				0	30	60	90	120	150	180
100	0	0	450	2950	1980	1920	1936	1943	1961	2215
90	10			2970	2708	2624	2633	2685	2691	2755
80	20			3100	3024	3020	3062	3081	3094	3181
70	30			3156	3150	2980	2996	3030	3051	3064
100	0	1		1820	300	141	162	180	210	250
90	10			2035	450	255	275	299	310	364
80	20			2375	650	360	368	390	431	465
70	30			2916	860	510	516	542	572	600
100	0	0	250	2950	1980	1920	1936	1943	1961	2215
90	10			2963	2711	2585	2589	2593	2602	2641
80	20			3002	2981	2783	2810	2826	2851	2916
70	30			3114	3044	2862	2911	2918	2962	3000
100	0	1		1820	300	141	162	180	210	250
90	10			1910	320	290	300	310	320	470
80	20			2030	400	360	370	378	430	508
70	30			2250	514	444	446	453	494	660

Comparing the data on the minimum effective viscosity of suspensions without the addition of C-3 and with its addition in an amount of 1% of the binder weight, we see that high-speed mixing in the presence of the superplasticizer C-3 leads to a decrease in $\eta_{eff.}$ in comparison with the control (the suspension was not subjected to activation; C-3 = 0%) from 2950 to 141 cP, that is, more than 20 times. The graphical dependences shown in Fig. 1 indicate that the time of mechanical activation of the cement suspension has a great influence on the decrease in the effective viscosity.

$\eta_{eff,}$ cP

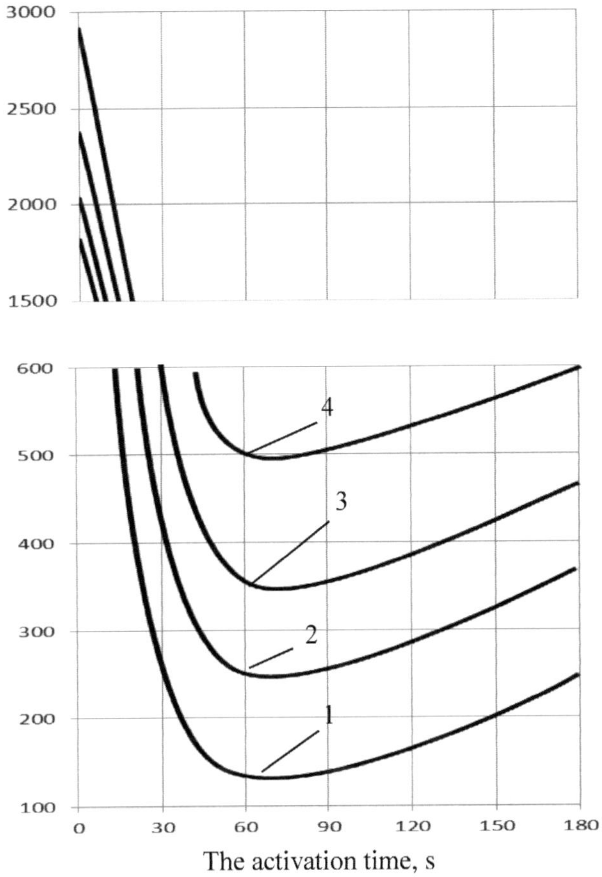

Figure 1. Influence of the activation time of cement-containing suspensions on the change in its effective viscosity. Specific surface area chamot = 250 m^2 / kg 1,2,3,4 content of ground chamot, respectively 0%,10%, 20% and 30%

The greatest decrease in viscosity is observed in the first 30 seconds of high-speed mixing of the suspension. Having reached the minimum value, the effective viscosity of the suspension begins to increase, reaching a maximum after 180 seconds of activation. In our opinion, this can be explained by the fact that in the process of intense hydrodynamic effects on cement-water systems in them, along with the destruction of spatial aggregates and the weakening of bonds between particles, the reverse process also occurs - thixotropic restoration of the structure, i.e., the formation of new aggregates and bonds.

The introduction of ground chamot into Portland cement leads to an increase in the minimum effective viscosity of the suspension achieved after 60 seconds of activation. So, at a 10% content of ground chamot in Portland cement, the minimum effective viscosity increases from 141 to 255 cP.

The introduction of 30% ground chamot causes a further increase in the effective viscosity of the suspension up to 510 cP. It has been experimentally established that the specific surface area of ground chamot also affects the change in the effective viscosity of the suspension. Comparing the values of its effective viscosity after 60 seconds of activation, we see that as the specific surface of the chamot increases, η_{eff} increases. So, if η_{eff} for Portland cement with the addition of 30% ground chamot with S = 250 m^2 / kg is characterized by an effective viscosity of 444 cP, then the introduction of the same amount of chamot into Portland cement with S_{sp}= 450 m^2 / kg causes an increase in the effective viscosity to 510 cP. Thus, it can be stated that the introduction of ground chamot into Portland cement increases the effective viscosity of the suspension both on a mechanically activated binder and on a binder that has not been mechanically activated. An increase in the specific surface area of chamot leads to an increase in effective viscosity by an average of 20%.

To assess the effect of high-speed mixing in the presence of C-3 superplasticizer on the cement-containing suspension, a dimensionless coefficient (K) was chosen, which was defined as the ratio of the viscosity of a practically intact (η_0) structure to the minimum possible viscosity that the suspension acquires as a result of the application of mechanical activation (η_{ma}), the introduction superplasticizer C-3 ($\eta_{C\text{-}3}$), or their combined effect (η_{Σ}^{e}):

$$K_{ma} = \frac{\eta_0}{\eta_{ma}} \quad (1)$$

$$K_{c-3} = \frac{\eta_0}{\eta_{c-3}} \quad (2)$$

$$K_{\Sigma}^{e} = \frac{\eta_0}{\eta_{\Sigma}} \quad (3)$$

The calculated coefficient of effective viscosity reduction under the combined action of high-speed mixing and superplasticizer C-3 (K_{Σ}^{c}) on the suspension is a function that depends on the effectiveness of each of these factors separately:

$$\eta_{\Sigma}^{c} = \eta_0 \frac{\eta_{ma}}{\eta_0} \cdot \frac{\eta_{c-3}}{\eta_0} \quad (4)$$

or $$K_{\Sigma}^{c} = \frac{\eta_0}{\eta_{\Sigma}^{c}} = K_{ma} \cdot K_{C-3} \quad (5)$$

The level of synergy (Y_c) was determined as the ratio of the actual coefficient of viscosity decrease (K_{Σ}^{e}) obtained experimentally to its calculated value:

$$Y_s = \frac{K_{\Sigma}^{e}}{K_{\Sigma}^{c}} \quad (6)$$

The obtained results of experimental studies are presented in table. 2.

Table 2. Synergy level

Chamot, %	Superplasticizer C-3, %	Y_s		
		Specific surface area, m^2/kg		
		250	350	450
0	1	4.49		
10		6.38	6.44	6.50
20		6.4	6.57	6.62
30		4.67	4.85	5.35
0	0.5	3.8		
10		4.37	4.42	4.46
20		4.5	4.77	5.1
30		4.36	4.56	4.68

It was found that the level of synergy depends on the concentration of C-3, as well as the concentration and specific surface area of ground chamot.

The greatest influence on the level of synergism is exerted by the concentration of the reducing additive C-3. The level of synergism with simultaneous action on the suspension of high-speed mixing and the addition of C-3 in an amount of 0.5% increases, as compared to the level of synergism of the suspension with the addition of C-3 = 1% from 3.8 to 4.94, i.e. 23% higher.

When 10% ground chamot is added to Portland cement, the level of synergism increases to a value of 6.44. A further increase in the content of chamot in Portland cement slows down the growth of the level of synergy, and at a 20% content of chamot in the binder, its value is 6.57.

A further increase in the content of ground chamot in Portland cement leads to a decrease in the level of synergy at 30% of its content in Portland cement, the value of Y_s is 4.85. It should be noted the effect of the specific surface area of chamot on Y_s - with an increase in $S_{sp.}$ from 250 to 450 m^2/kg, the level of synergy increases from 4.36 to 4.68, that is, by 7%.

5. Conclusions

Activation of a cement suspension in a high-speed tribomixer and in the presence of C-3 superplasticizer leads to a sharp decrease (20 times) in its effective viscosity in comparison with the control.

The introduction of ground chamot into the binder leads to an increase in the minimum possible effective viscosity of the cement slurry from 141 cP (the content of ground chamot in Portland cement is 0%) to 444 cP (the content of ground chamot in Portland cement is 30%).

References

[1] Hasan N Light concrete on a new composite binder 2004 *Bulletin of Vinnytsa Polytechnic Institute* **6** p 10-15

[2] Barabash I 2002 *Mechanochemical Activation of Mineral Weavers* Odessa Astroprint p 100

[3] Fedorkin S 1997 *Mechanoactivation of secondary raw materials in the production of building materials* Simferopol Tavria p 180

[4] Ksonshkevych L 2012 *High-strength concrete on the mechanically activated binder* Dis. Cand. tech Sciences: 05.23.05 Odessa p 145

[5] Ksonshkevych L, Krantovska O, Petrov M, Synii S and Uhl A 2018 Investigation of the structure of cement stone, obtaining and optimization of high-strength concrete on mechanically activated binder *Reliability and Durability of Railway Transport Engineering Structures and Buildings:* MATEC Web Conference https://doi.org/10.1051/ matecconf/201823003010

[6] Ksonshkevych L, Barabash I, Krantovska O, Synii S, Sunak P 2019 Disperse reinforced concrete with polycarboxylate additive on a mechanically activated binder *Materials Science and Engineering:* IOP Conference Series https://iopscience.iop.org/ article/10.1088/ 1757-899X/708/1/ 012092

[7] Babkina L, Soloshenko L, Karjakina E, Tishina T, Shcherbak L 2014 Study of the influence of alumina cements from different manufacturers on the properties of chamot concrete mixture and samples from it Collection *of Science Practices PAT "UKRNDI VOGNETRIVIV IM. A. S. BEREZHNOGO"* **114** p 67-75

[8] *Primachenko* V, Babkina L, Soloshenko L 2012 Investigation of the influence of the type of hydraulic binder on the properties of chamot concrete and samples from it Collection *of Science Practices PAT "UKRNDI VOGNETRIVIV IM. A. S. BEREZHNOGO"* **112** p 184-190

[9] Bazhenov Y 2011 *Concrete technology* Moscow Publishing ACB p 528

Triad «Structure, Information, Fractality»

V Martynov[1]

[1] Department of Production of building products and structures, Odessa State Academy of Civil Engineering and Architecture, Didrikhsona st., 4, 65029, Odessa, Ukraine ORCID: http://orcid.org/0000-0002-9674-7920

E-mail: ogasa_psk@ukr.net

Abstract. It is focused on the point that, factually, the material's structure is a primary cause of its properties. The elements of hardened cement stone (inner interfaces), which are the source of its properties, are revealed. It is supposed that the inner interfaces branched chain (patterns) are carriers of certain information. It is proposed to describe this information in quantitative values by fractal dimensions. With such approach the research process is described by sole triad «structure-information-fractality». This article states experimental results proving the effectiveness of the approach proposed.

Constructional materials properties define their relevant functional purpose in the field of construction. Production of composite building materials with given properties is carried out by sequence of technological operations in conformity to the scheme: «composition - technology - property» [1]. For the purpose to determine material stress-strain properties it was developed many tools, devices, ways and methods. One of the basic research methods used with this scheme is a pattern «black box». It was suggested by Ross Ashley, one of the founders of cybernetics [2]. The drawback of this approach is that the research process does not specify the reason of the result obtained. At the same time, contemporary research community have no doubt that materials properties in chains of cause-effect relationship, are attributed, in the first turn, to the character of their structure [3,4]. In this case the above stated chain will feature one more element «structure». Exactly a structure character is a primary cause of the properties of the material obtained. In studying of the material, a researcher becomes a participant in «subject - object» relationship. In this case material and its structure will be «object» of the research. That means that the structure of the material is objective factor, and the research is of subjective nature. Effectiveness and validity of their results depend on many factors: what scientific paradigm is observed by researcher, what methods and tools he uses etc., in other words, what volume and quality of information he has.

The special meaning was attached to the term "information" in XX century. Information (informatio - informing, explanation, narration) is, in a broad sense, an abstract concept, which has many meanings depending on a context [5], and, in its narrow sense, - findings (data, messages) irrespective of form of their representation. Information determines properties of objects and phenomena in space and time. Thus, the research object is informational but cannot convey it to you directly. One of the methods due to which a researcher can obtain this information is disclosing of material structure.

For this purpose, in the first turn, it is necessary to define the term of "structure". Let's consider it by way of example of" cement stone", which is base material for various kinds of concrete. The hydration of cement is accompanied by reduction of the total volume of the system [6]. Eventually, after doing so, the cement stone will develop a branched chain of inner faces of inner interfaces (II).

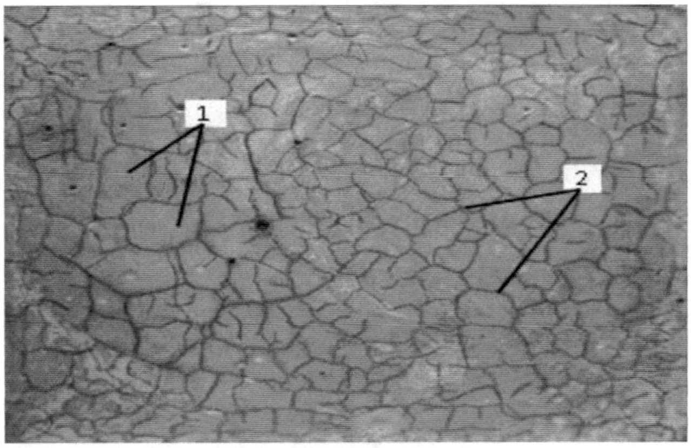

Pic.1. Cement stone structure:
1 – structural elements (units);
2 – border-lines (inner interfaces) between elements.

II are located between separate solid-phase units (elements). This is the ground to consider the cement stone structure as a sum total of interrelated elements making the sole unit. In such representation the cement stone structure meets all the features of the concept «system». Its separate solid-phase units are the elements of the structure, and inner interfaces serve as connections between the elements. These connections ensure the integrity of the object, which is the main definition of any system, and the terms "structure" and system become identical. Besides, in studying of the research object one will be able to see the ground for application of systemic laws, in this particular case it means bringing the object properties into correlation with qualitative and quantitative links' characteristics, namely II. In other words, the more will be a total inner interface length, the better endurance the material will have.

As you can see on picture1, II makes various figures, patterns in cement stone. In terms of physics, this structure represents percolating cluster [7]. There are grounds to suggest that the patterns are particular for each structure. We can also suggest that the patterns are carriers of some piece of information. However, it appeared to be complicated to decode this information. The task also becomes more difficult due to the fact that, in terms of classical mathematics and geometry, it is impossible to describe their shape due to their "curvedness" , to do it we would need some translating program.

The first scientist who noticed that there are no straight lines, "correct" geometrical figures in nature etc. was Mandelbrot. He became the author of Fractal Geometry and introduced the term "fractal" (fractionary) [8]. To make a qualitative assessment of the line curvedness grade (deviation from directness), wall and surfaces roughness, Mandelbrot suggested using of fractionary dimensions (Hausdorff-Besikovitch dimensions) – the dimensions which differs from the whole-numbered topological dimensions (0, 1, 2, 3), for example - 1,56. This number is an index of grade of line deviation from its directness, that means this is not a direct line, but also not a flat figure, since the line is not closed. The more the number approximates to the digit 2, the more curved the line will be. By using of mathematical methods, mathematicians made many various fractals. The most famous among them are: "Koch snowflake", "Cantor set", "Mandelbrot set", "Sierpinsky carpet", "Menger sponge" etc.

The other, not less important fractal property is self-similarity. The simplest case is a fractal the small part of which contains the information about the whole fractal. Starting his research exclusively from mathematical operations, eventually Mandelbrot came to philosophical conclusion к философскому on a fractal nature of the whole world. Everything is structured according to the fractal principle. This concerns both shape, and contents of the informational objects and phenomena of the world. Fractals can describe with high level of preciseness many physical phenomena of the real world which cannot be described using simple geometrical figures. For example, shapes of mountains, clouds, roots, branches and leafs of the trees, blood vessels, patterns of population growth, shifting from orderliness to chaos etc.

Many theoretical fractals can serve as patterns for composite building materials. For example, «Menger sponge» is a pattern for cellular concrete.

Thus, if a cement stone structure as patterns from inner interfaces is a carrier of information, which may be interpreted in terms of fractal geometry, then the logical triad возникает логическая триада «structure-information-fractality» will appear. In common terms, triad is understood as a unity made by three separate members/parts. There are many triads, for example, Hegel triad Гегеля. This is an association of any opposite concepts and any third concept, which mediates (that is, shows) an internal unity of two opposite concepts [9]. In our case the concepts «structure» and «information» cannot be considered as exact antithesis. At the same time, proceeding from the definitions of these concepts, it is also hard to find out any similar signs of them. The introduced third concept "fractality" is exactly the element of the triad which allows us to show the internal unity of the first two concepts.

The Department of Production of Building Products and Construction of Odessa State Academy of Civil Engineering and Architecture have developed computer programs on determining of fractal dimensions of building composites structure.

In the earlier research [9] it was discovered that the strength of macroporous materials (cellular concretes) is not directly proportional to the strength of solution component that is a base material for structural support. At the same time the strength of construction dense composites on mineral cementitious matters is directly proportional to cement-water ratio. The table 1 states the results of experiment on determining properties of non-autoclaved cellular concrete and dense solution component, from which this cellular concrete was produced. The experiment involved changing of rheological conditions of obtaining cellular concrete due to water requirement of mortar mix, which was measured by the diameter of spread of mortar mix throughout Suttard's viscosimeter. After preparing of solution component of the given water requirement its certain part was selected and it was used for shaping of prism samples, with the dimensions 4x40x160 mm. Foam was added to the rest part of the solution in the amount required for obtaining cellular concrete of average density of 600 kg/m^3 in dry state. The samples were tested in the age of 28 days.

Table 1. Cellular concrete and solution component properties

Properties	Diameter of solution spread , mm				
	200	240	280	320	360
Water-cement ratio	0,5	0,59	0,63	0,67	0,69
Average density, kg/m^3	620	618	612	608	615
Humidity, %	25,3	24,4	23,7	27,9	28,7
Cellular concrete strength, MPa	0,2	0,9	1	1,1	1,3
Solution component strength, MPa	36,7	34	28,4	25,6	24,3

The above stated results prove the hypothesis that if the water-cement ratio increase the solution component (solid) strength will decrease and cellular concrete strength will decrease. In this connection it was supposed that the properties of porous materials are determined by the nature of solid phase spreads (interporous parting walls), i.e. by the nature of structure [10].

In order to study the structure formation of materials and porous structure an experiment which modelled various environmental rheological conditions was conducted. The study was carried out using pattern material. As an analogue to cement paste, water-slurry mortar with various water-slurry ratio was used. Besides, the dimensions of pores in the porous materials patterns were changed. Patterns were kept in closed premises at the temperature of 22±2°C. When removing free water (drying out), the changes in patterns appearance were observed. At the first stage separate fractures appeared in the continuous medium of dense-structured materials' patterns. Further, on its drying out, the fractures spread and closed on each other finally making chain of inner interfaces. At first pore simulators were inserted into the of porous materials pattern, and after reaching certain strength by the material patterns were extracted. Formation of inner interfaces ran in interporous walls. The final view of the patterns you can see at the picture 2.

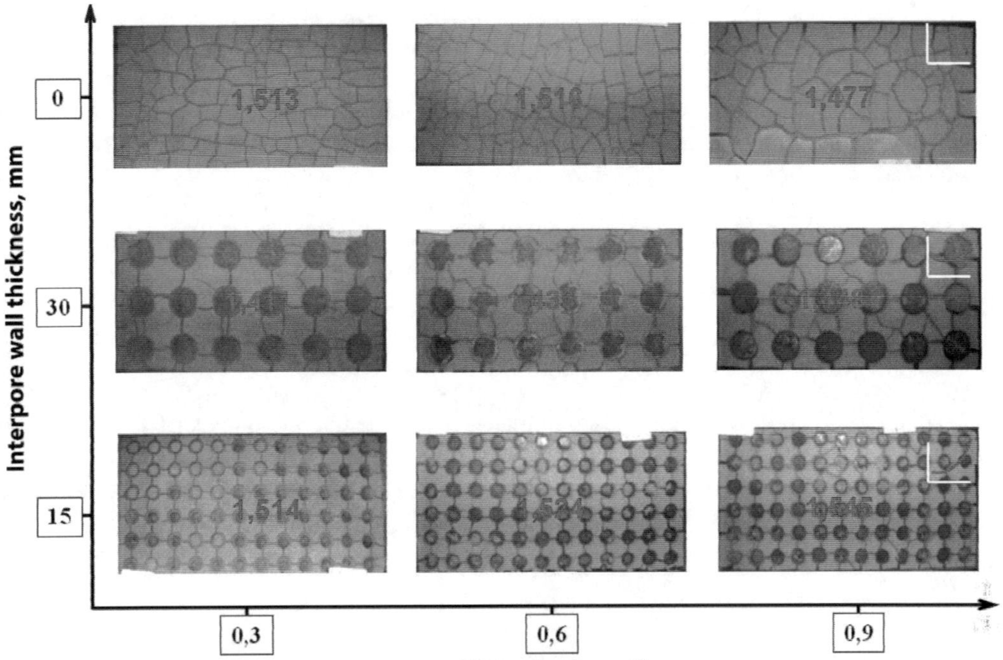

Pic. 2 Physical patterns of composite building materials

In the center of the patterns you can see fractal dimensions' values. In the right upper corner there is a graph of change of fractal dimensions depending on water-cement ratio. As it is observed, in dense-structured materials patterns (upper row) increase of water-cement ratio results in decrease in fractal dimensions фрактальной размерности уменьшаются, and in porous material patterns -vice versa. Besides, a fractal dimensions' value depends on pores dimensions: the smaller pores, the more is fractal dimensions' value.

Thus, it allows to establish a direct informational cause-effect link between a fractal nature of a structure and physical-mechanical properties of materials, which, in its turn, provides possibilities of artificial making of materials with the given structure and synthesis of the material with prognosable properties.

Resume

1. Physical-chemical processes accompanying the hardening process in the cement stone results in formation of a branched chain of inner interface as various patterns. Inner interfaces are represented as the structure elements, on nature of spreads of which the cement stone properties depend.
2. It is supposed that II patterns, being characteristics of structure are the carriers of certain information, which is supposed to be assessed in quantitative values using concepts and methods of fractal geometry. Fractality, being a universal quality of nature, is an element of the triad «structure-information-fractality» manifesting internal unity of these first two. Application and implementation of the proposed triad open the possibilities for artificial making of materials with the pre-set structure and synthesis of the materials with prognosable properties.
3. The article states the results of experimental studies which proved the effectiveness of the proposed approach. The experimental studies allowed us to explain various water-cement ratio influence on the change of strength of materials of dense and porous structure.

References

[1] J R Hattrick-Simpers, J M Gregoire, A Gilad Kusne 2016 Perspective: Composition–structure–property mapping in high-throughput experiments: Turning data into knowledge *APL Materials* **4** 053211
[2] Ashby, W Ross 1956) *An Introduction to Cybernetics* London Chapman & Hall Ltd

[3] Vyrovoy V, Korobko O, Ielkin O 2017 *Technological events in structural evolution of building composites* Web of Conferences Open Access Journal «MATEC Web of Conferences» **16** doi: 10.1051/matecconf/201711601021 p 6
[4] Vyrovoy V, Korobko O, Sukhanov V, Posternak O 2019 Concrete as a self-organizing system *IOP Conference Series: Materials Science and Engineering. Reliability and Durability of Railway Transport Engineering Structures and Buildings* **708** p 6
[5] https://dl.nure.ua/pluginfile.php/
[6] S Rahimi-Aghdam, E Masoero, M Rasoolinejad & Z P Bažant 2019 Century-long expansion of hydrating cement counteracting concrete shrinkage due to humidity drop from self-desiccation or external drying *Materials and Structures* **52** Article number 11
[7] Christensen K 2002 Percolation Theory *Dr. Blackett Laboratory Imperial College London*
[8] Mandelbrot B 1983 The fractal geometry of nature N. *Y. Freeman* p 480
[9] Martynov V, Vyrovoi V, Vetoh A, Martynova E, Elkin V 2014 Modeling in Assessing the Nature of the Structure of Foam. Building Materials Equipment Technologies of the XXI century *Information Scientific and Technical Journal* Moscow **11(190)** p 11-13
[10] Martynov V and others 2015 Influence of the Structure of a Material Solid Phase on the Properties of Cellular Concrete *National Journal of Composite Materials №4* **5** p 79–80

Evaluation of the efficiency of using solar energy in the production of building composites

A Plit[1], M Zavoloka[1], O Shynkevych[1], V Vinnichenko[2]
[1]Odessa State Academy of Civil Engineering and Architecture, Didrikhsona st., 4, 65029, Odessa, Ukraine
[2]Kharkiv National University of Civil Engineering and Architecture, Kharkiv, Ukraine

E-mail: alex89805@gmail.com

Abstract. The results of research of speed of set of durability of fine-grained concrete are in-process presented on a compression which hardened in solar chamber of cyclic action. The analysis of existent types of industrial solar chambers is conducted and basic structural elements are considered. Resulted calculation of thermal balance of solar chambers. Experiments are conducted on compositions of fine-grained concrete on cement astringent with the use of additions of superplasticizers of firm MC - Bauchemie: MC - PowerFlow 3100 and MC - PowerFlow 2695. After self-control in solar chamber of cyclic action durability was determined determined at the different terms of hardening. The comparative analysis of results of durability is conducted on the compression of concrete in solar chamber and chamber of the normal hardening.

1. Introduction

The analysis of literary data showed that industry of the collapsible reinforced concrete is the large consumer of thermal energy. The most power-hungry process in this area is a technological process of thermal treatment of wares, which more than 70% energy is expended on. About 90% of general volume of reinforce-concrete wares (RFC) is exposed to treatment steam. On the receipt of steam the unprocessed in natural energy sources is spent. Taking into account the growing increase of their cost, heat and moisture treatment it is necessary to acknowledge inefficient in a power relation. In this connection the problem of the rational use of traditional fuel and energy resources and plugging in power balance of untraditional sources of energy becomes all sharper, such, as energy of a sun, winds, deep warmth of earth and secondary warmth of various productive processes. Presently there is a tendency of increase of scales of the use of proceeded in energy sources and foremost by a sun. Application of sun energy finds the application in the various areas of productive processes, that especially topically for the south areas of Ukraine [1-3]. The large variety of the solar radiation engineering systems thermal treatment of building compos can be carried out in which is known. Thermal options, different inverter devices, sun batteries, behave to them.

From data of the World power agency [4] sun energy sources to 2019 attained 9% of world electro-generation, and to 2040 must make 24%. It is possible only on condition of speed-up development and mastering of untraditional sources energies which sun energy belongs to the number. All types of energy, producible on the surface of planet, in the end transformed in a warmth and heat an atmosphere. Before humanity there is a new problem of "thermal contamination" of air pool of planet.

A sun radiation, coming on Earth, does not change thermal balance of planet and is the "clean" type of energy.

Use of sun energy for thermal treatment of concrete in daylight saving time and, foremost, in districts with a dry hot climate allows considerably to shorten labour intensiveness and terms of humidity care of concrete, and consequently, and terms of handing over of monolithic constructions under loading as a result of speed-up set of durability at enhanceable temperatures, practically assuredly to provide the required quality of concrete, his longevity and firmness to atmospheric influences as a result of the rapid physical and chemical fastening of water, management the mode of heating and maintenance of more homogeneous temperature field in a construction in the process of all thermal treatment.

In the world certain experience is already accumulated in area of the use of sun energy in a national economy. Worked out and the different industrial and domestic solar energy systems and options are used. For providing of safety of surface of concrete different film-forming suspensions and composition compositions are worked out, including. hydrophobic or as self-destructive suds and another ways of care of concrete with the use of polymeric tapes for defense of surface arcwise extensive concrete constructions [5,6].

Solar energy treatment is warming up of concrete direct influence of solar radiation through translucent coverages or sun receivers. The simplest method of the use of sun energy for warming up of concrete is application of helioform, which consist of two basic elements: actually forms and lids with translucent heat-insulation coverage or in forms with the use of moisture-proof perceiving sun coverage in combination with thermostat self-control. At thermal treatment 0,6 ... 0,75 makes in solar chambers or solar chambers coefficient of the useful use of sun energy. Except for it, the stream of sun energy can influence on solar thermal receiver, coolant (water, butter of and other) circulates in which, that heats thermoforms or thermal compartments additionally. Polymeric photo-electric elements are promising coverage for refractions of visible spectrum. As spectrums of absorption organic polymeric and low-molecular types of semiconductors, are not continuous, as in inorganic semiconductors. As a result, a construction from organic materials is able to skip visible light and take in an invisible spectrum, for example infra-red. According to power distribution of sun spectrum the more than half of sunlight is distributed in an infra-red area. Consequently, theoretical efficiency of polymeric photo-electric elements with absorption only of infrared can be the same high, as at a device with absorption only of visible light. (7-9).

On the first stage of researches experimental works are conducted on determination to influence of solar chambers on a cement matrix for the concretes of reinforce-concrete wares and arblite concrete under influence of additions of superplasticizers of the German firm MC - Bauchemie.

On the second stage of research a calculation is executed on determination to the amount of energy, necessary for heating of wares in solar chambers. Main part of helioforms are sun elements, which are semiconductor photo-electric elements, intended for direct transformation of sun radiation to electric energy. They are included in more wide range of devices of photocells. Usually sun elements have a large area of receiving surface and major parameters for them are temperature stability, coefficient of transformation of light energy in electric one, radiation firmness and others. Mostly sun elements are the constructions from semiconductor materials, well absorptive a sun radiation, metallic and dielectric layers. If the device contains electrolyte solutions, they are called photoelectrochemical converters.

The functioning of solid-state solar cells is based on the phenomenon of the photoelectric effect of the valve. The eaten up by a semiconductor light quanta release electrons and holes from intracrystalline connections and transfer them in the mobile state. Whereupon the oppositely charged free charge carriers are spatially divided and move in opposite directions in investigation of diffusion or drift in the power field which initially must be in a semiconductor [10].

The most essential task for sun elements is an increase of their efficiency. Efficiency of work of photocell increases at the use of the concentrated sunlight, however it is here necessary to delete a surplus heat from a photo consumer. To date the highest index efficiency, the scientists of the National laboratory attained on the study of proceeded in energy (THE USA) and attained a mark in 47% [11], by the concentrated sun light, setting thus a new pestilent record for sun batteries. Among industrial prototypes shows the best indexes elements on the basis of silicon, although with the mass production of high-efficiency and expensive sun elements economic ineffective elements compete from more cheap explorers. Industrial sun batteries work with the use of different sun elements, basic from which are silicic, single-crystal, polycrystalline, amorphous pellicle, polymeric, photo sensitized, which differ on efficiency and cost. At development of effective elements of effort, it is necessary to send not only to their efficiency but also on the compactness of the systems. It is needed to pay attention to plastic crystals due to which it is possible to promote the compactness of the systems and shorten expenses on their production (12-15).

Advantages of solar thermal treatment:

- it is an economy of unprocessed in fuel and energy resources,
- it is application ecologically of clean thermal energy source,
- it is a decline of prime price of concretes,
- it is diminishing of capital costs on a boiler room, heating mains of and other,
- allows to yield up to traditional thermal treatment in warm periods of year in districts to the south of a 50° north breadth.

For heating of concrete, fine-grained concrete and arblite concrete wares a calculation is executed on determining the necessary amount of energy.

The methods of calculation consist of three parts:

- it is a calculation of thermal balance of chamber heat treatment is carried out in which. as a result of calculation, the amount of energy, necessary for heating is determined,
- to define the amount of heat which can be got a sun collector. Such data for the set geographical co-ordinates are presented NASA. For the basic cities of Ukraine information of presented in the Internet [16],
- to find the area of sun collectors which will provide the necessary amount of energy. Equalization of thermal balance has a next kind [17]:

$$Q = Q_1 + Q_2 + Q_3 + Q_4 + Q_5 + Q_6 + Q_7 + Q_8 + Q_9 + Q_{10} + Q_{11} - Q_{12} \qquad (1)$$

where Q_1 is an expense of heat on heating of components of raw material mixture of concrete, Q_2 is an expense of heat on heating of shutting water, Q_3 is an expense of heat on evaporation of part of water, Q_4 is an expense of heat on heating of armature, Q_5 are expenses of heat on heating of forms, Q_6 - warm, which is accumulated by the walls of chamber, Q_7 is an expense of heat on heating of lid of chamber, Q_8 is an expense of heat on heating of free space in a chamber, Q_9 are defervescence due to the loss of heat through unclosenesses, Q_{10} are defervescences in an environment the walls of chamber, Q_{11} are defervescences in an environment the lid of chamber, Q_{12} is a heat of exothermic reactions, flowing in a concrete, Q is an amount of energy, necessary for heating, kj / cycle.

Amount of heat which can be got a sun collector [18,19]:

$$Эк = E \cdot S \cdot \eta \qquad (2)$$

where E is a middle daily level of solar radiation, kW o'clock / of m^2 / day, S is a working area of sun collector, m^2, η is an output-input of sun collector ratio.

Area of sun collectors which will provide the necessary amount of energy, m^2.

$$S = Q / E\eta \qquad (3)$$

2. The methods of realization of experiment consisted in the following: the laboratory model of solar chambers of cyclic action of type was set the "Hot box". The construction of solar chambers consisted of wooden framework on which polyethylene tape was strained in two layers, with a gap between layers 2 sm. A size was appointed so that to cover all threefold forms with concrete mixture. Forms with concrete mixture were placed in solar chambers and maintained there to testing. The coulisse of standards for tests was produced on 1, 3, 7, 14, 28 twenty-four hours. For the investigated compositions in parallel control standards-were made twins which was kept in the chamber of the normal hardening with a temperature 20±3°C and by humidity 95±5 %. Making of standards of beams was produced in accordance with DSTY B V. 2.7-214:2009 [20]. Tested on an attorney equipment of

Ukraine accredited by national Standard scientifically to the research laboratory. Durability was measured on a 100 ton press on a minimum range with the maximal loading in 20 t and scale of division in 40 kg. Treatment of results passed in the program Excel on calculations in accordance with DSTY B V. 2.7-214:2009, cited data as middle arithmetic from all results of one party of standards.

3. Basic material

Results are presented of the conducted experiment on speed of set of durability of standards of cement compos, kept in solar chamber for comparing to the results of control standards, which was kept in the standard chamber of the normal hardening. For comparison standards were made with the use of two additions of firm MC - Bauchemie: by complex addition accelerating of hardening of MC - PowerFlow 3100 and superplasticizer of MC - PowerFlow 2695.

Control standards are kept in the chamber of the normal hardening collected vacation durability of 70-80%, the got durability provides removal of planking on a 7 twenty-four hours, for standards which was made with the use of addition of MC - PowerFlow 3100. At the use of addition of MC - PowerFlow 2695 vacation durability was attained on a 14 twenty-four hours. The results of realization of strength tests are driven to the table. 1 and on a fig. 1.

Table 1. Results of tests of concrete at storage in the chamber of the normal hardening

Name	W/B	Twenty-four hours	Tensile strength on a compression, MPa	Name	W/B	Twenty-four hours	Tensile strength on a compression, MPa
MC - PowerFlow 3100	0,3	1	12,6	MC - PowerFlow 2695	0,32	1	1,1
		3	27,8			3	5,1
		7	35,4			7	11,8
		14	41,6			14	16,3
		28	49,9			28	26,5

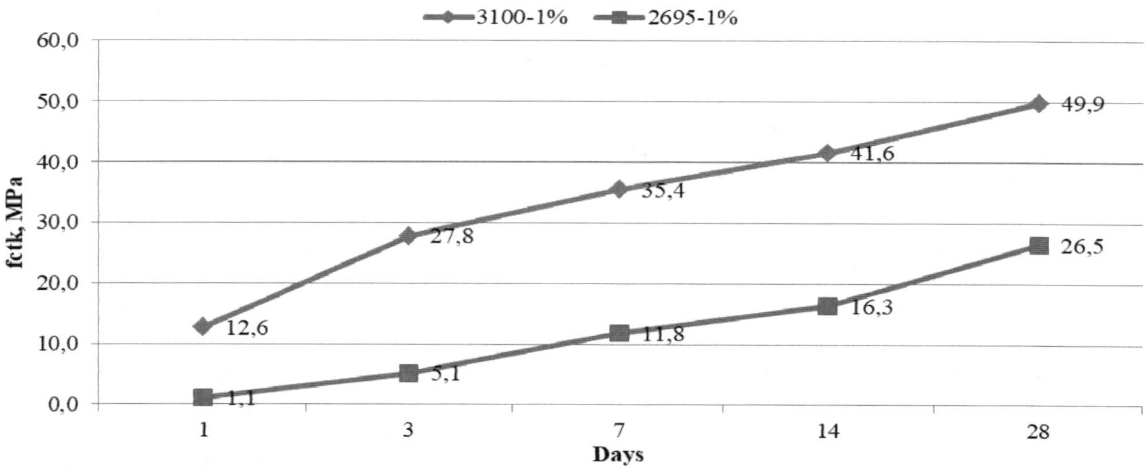

Figure. 1. Dependence of durability on the compression of concrete from his age, at the use of addition of MC - PowerFlow 3100 and MC - PowerFlow 2695, at storage in the chamber of the normal hardening

In solar chamber due to the solar thermal warming up of standards the accelerations of set of durability of concrete are provided. Vacation durability in 70-80% attained in the flow of a 3 twenty-four hours for a standard with the use of addition of MC - PowerFlow 3100. In also time standards, for making of which addition of MC was used - PowerFlow 2695 collected the same percent of durability arrived at after a 7 twenty-four hours of hardening it is presented on a fig. 2.

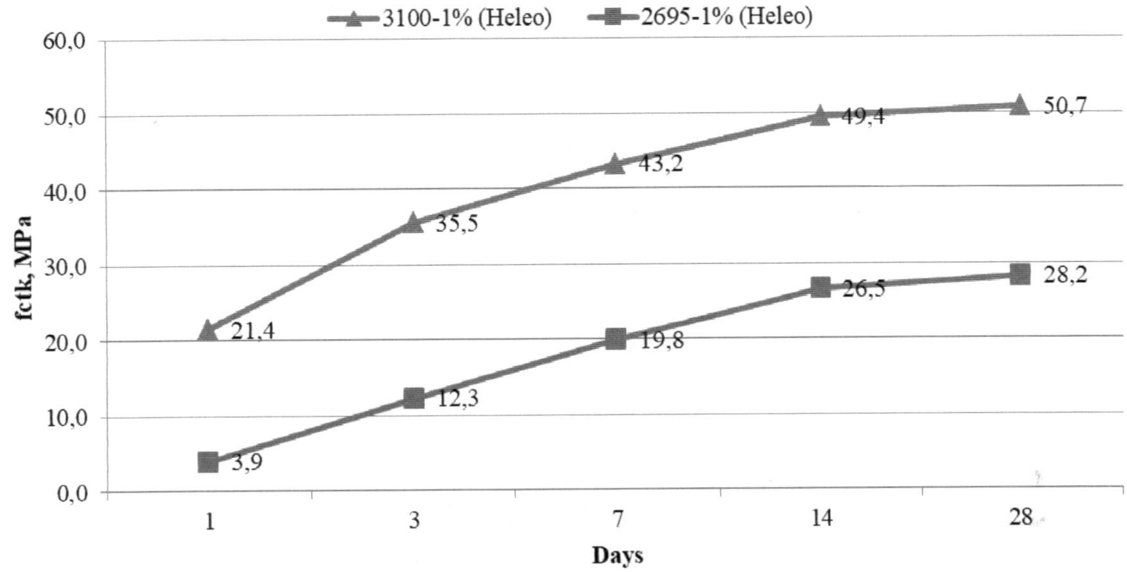

Figure. 2. Dependence of durability on the compression of concrete, at storage in solar chamber from his age, at the use of additions of MC - PowerFlow 3100 and MC - PowerFlow 2695

Comparison of durability was also conducted between standards from solar chamber and chamber of the normal hardening. Standards, collecting durability in solar chamber showed the speed-up set of durability, at equal terms of hardening. Results are driven to the table. 2.

Table 2. Results of tests of concrete at storage in solar chamber

Name	W/B	Twenty-four hours	Tensile strength on a compression, MPa	Increase of durability on a compression in solar chamber, %
MC - PowerFlow 3100	0,3	1	21,4	169,8
		3	35,5	127,7
		7	43,2	122,0
		14	49,4	118,8
		28	50,7	101,6
MC - PowerFlow 2695	0,32	1	3,9	354,5
		3	12,3	241,2
		7	19,8	167,8
		14	26,5	162,6
		28	28,2	106,4

During realization of experiments strength descriptions of concrete mixture were set. Standards are maintained in solar chamber on the first twenty-four hours showed durability higher on 254,5%, what analogical standards from the chamber of the normal hardening. On a 3 twenty-four hours of storage a maximal increase made 141,2%, on a 7 twenty-four hours of 67,8%, on a 14 twenty-four hours of 62,2% and on a 28 twenty-four hours of 6,4%.

Experiments were conducted in springtime of year, which is not the hottest, and solar chambers works fully on sun energy and fully depends on an ambient temperature. For determination of efficiency of work of solar chamber thermometers were set taking temperature environment, to concrete mixture and temperature inwardly solar chambers. In the process of realization of experiment, the maximal values of temperature were educed: environment - 38 °C, solar chambers - 46 °C and concrete mixture 52 °C.

Basic conclusions

Experiments, conducted on two additions superplasticizers of production of firm MC, - Bauchemie, showed that application of solar chamber improved the primary set of durability considerably, and helps in the earliest possible dates to collect vacation durability, that accelerates building in turn. Also on a 3 twenty-four hours of self-control standards with addition of MC - PowerFlow 2695 showed to 43% durability, and standards with addition of MC - PowerFlow 3100 showed to 70% durability. The variant of calculation of solar chamber is in-process presented which can be applied for further perfection of existing and creation new effective solar chambers and is one of directions of development of production of concrete wares in districts with a hot climate.

References

[1] Malyarenco V 2004 *Energy of the environment* Energy Saving (Kharkiv: Rubikon)
[2] Alaa A F Husain, Wan Zuha W Hasan, S Shafie, M N Hamidon, S Pandey 2018 A review of transparent solar photovoltaic technologies. In Renewable and Sustainable Energy Reviews **94** pp 779-791 (DOI: https://doi.org/10.1016/j.rser.2018.06.031)
[3] Esence T, Benoit H, Poncin D, Tessonneaud M, Flamant G 2020 A shallow cross - flow fluidized - bed solar reactor for continuous calcination processes. In Solar Energy **196** pp 389-398 (DOI: https://doi.org/10.1016/j.solener.2019.12.029)
[4] *International Energy Agency [online] [accesat 26.03.2021] https://www.iea.org/data-and-statistics/?country=WORLD&fuel=Renewables%20and%20waste&indicator=SolarGen*
[5] Johnson J M-F, Reicosky D, Allmaras R, Archer D and Wilhelm W 2006 A matter of balance: Conservation and renewable energy. In Journal of Soil and Water Conservation **61(4)** 120A-125A; https://www.jswconline.org/content/ 61/4/120A
[6] Chang S-Y, Cheng P, Li G, Yang Y 2018 Transparent Polymer Photovoltaics for Solar Energy Harvesting and Beyond **2(6)** pp 1039-1054 (DOI: https://doi.org/10.1016/j.joule.2018.04.005)
[7] Geisz J F, France R M, Schulte K L, Steiner M A, Norman A G, Guthrey H L, Young M R, Song T, Moriarty T 2020 Six - junction III - V solar cells with 47.1% conversion efficiency under 143 Suns concentration. In Nature Energy **5** pp 326 - 335
[8] Serrano A, Durana M, Dauvergne J-L, Doppiu S, Palomo Del Barrioa E 2021 Tailored transition temperature plastic crystals with enhanced thermal energy storage capacity. In Solar Energy Materials and Solar Cells **220** (DOI: https://doi.org/10.1016/j.solmat.2020.110848)
[9] Zini E Green composites: an overview 2011 *Polymer Composites* **32** pp 1905-1915
[10] Geletyha G, Kudrya S 2005 Ukraine: unconventional and renewable energy sources Ukraine *In Zelena energia* **2** pp 8 - 10
[11] Mysak J, Voznyak O, Dacko O, Shapoval O 2014 Solar energy: theory and practice *Lviv Polytechnic Publishing House*
[12] Zhao Y, Procel P, Hana C, Mazzarella L, Tao G Y, Weeber A, Zeman M, Olindo I 2021 Design and optimization of hole collectors based on nc - SiOx: H for high - efficiency silicon heterojunction solar cells. In Solar Energy Materials and Solar Cells **219** (DOI: https://doi.org/10.1016/j.solmat.2020.110779)
[13] Montesa M J, Linares J I, Barbero R, Rovira A 2020 Proposal of a new design of source heat exchanger for the technical feasibility of solar thermal plants coupled to supercritical power cycles. In Solar Energy **211** pp1027-1041 (DOI: https://doi.org/10.1016/j.solener.2020.10.042)

[14] Yan L, Chouw N, Javaraman K 2014 Flax fibre and its composites *Engineering* **56** p 296-317
[15] Gulkis A 2009 The Energy Pyramid: The best path to lasting energy savings. In Journal of Soil and Water Conservation **64(5)** (DOI: https://doi.org/10.2489/jswc.64.5.143A)
[16] *Solar activity map in Ukraine [online]. [accesat 05.04.2021] https://www.solar-battery.com.ua/karta-solnechnoy-aktivnosti-v-ukraine/*
[17] Duffie J A, Beckman W A 1980 Solar engineering of thermal process *New York : John Wiley and Sons*
[18] Gopinathan K, Maliehe N, Mpholo M 2007 A study on the intercepted insolation as a function of slope and azimuth of the surface. In Energy **7** pp 213 - 220
[19] State standards of Ukraine B V.2.7-214 Construction Materials. Concrete. Methods for determining the strength of control samples (DOI:http://online.budstandart.com/ua/catalog/doc-page.html?id_doc=25943)
[20] Green Composites from Resources 2013 *CRC Press* p 419

Influence of matrix-filler thermal conductivity on micro heat transfer in two-component composites

A Pysarenko[1] **and I Zaginaylo**[1]

[1] Department of Physics, Institute of Civil Engineering, Odessa State Academy of Civil Engineering and Architecture, Didrikhsona st., 4, Odessa, 65029, Ukraine

E-mail: pysarenkoan@gmail.com

Abstract. The maps of local heat fluxes in the composite matrix with randomly located heat-insulating and heat-conducting inclusions have been calculated in this work. It was shown that with a significant difference in the thermal conductivity of inclusions and matrix, the main heat transfer occurs through the induced heat-conducting channels. It was found that the weighted average angles of local heat fluxes deviation from the direction of the temperature difference between the sample faces depend on the concentration of heat-insulating and heat-conducting inclusions of various sizes. A sublinear decrease or super linear increase in the effective thermal conductivity of the composite with an increase in the concentration of heat-insulating or heat-conducting inclusions, respectively, was associated with a change in the path length of local heat fluxes due to the flow at an angle to the direction of the temperature difference. An approximation formula for the dependence of the composite effective thermal conductivity on the filler concentration has been proposed, which makes it possible to take into account the presence of both heat-conducting and heat-insulating inclusions.

1. Introduction

The development of new, high-quality and cost-effective materials determines the development path of modern industry. The quantitative characteristics of new materials, as a rule, vary in wider ranges and, thus, correspond to the requirements for the operation of the finished product.

The possibility of changing the properties of composite materials, in particular the quantitative and qualitative characteristics of the filler, the physicochemical characteristics of the matrix, determines their competitiveness in relation to traditional materials. However, the economic efficiency of new materials largely depends on the ability to predict their basic integral and local characteristics, as well as operational properties. Great efforts are being made to improve the properties of composite materials and especially thermal conductivity. A wide range of changes in the thermal conductivity of a composite material determines a wide range of applications for these materials.

The concept of effective thermal conductivity makes it possible to determine in detail the behavior of composites from a technological point of view, which is an undoubted advantage of using composites in technological chains.

Now, there are a large number of theoretical and empirical models for predicting effective thermal conductivity [1 - 3]. However, despite the theoretical and technological importance of this property and a significant amount of research carried out, both the functional behavior and quantitative estimates of the value of composite effective thermal conductivity have not been sufficiently studied.

The reason for these difficulties is that the effective thermal conductivity is a complex function of the thermal conductivity of the filler and matrix [4], the geometry and spatial distribution of inclusions, filler-matrix thermal contact resistance [5]. The scope of analytical methods for solving problems of heat conduction comprises a fairly limited number of types of applications in which the geometry object is presented only simple shapes (circle, rectangle, fiber and so on).

An additional factor that complicates the theoretical description of heat transfer in composites is the random spatial distribution of the filler particles [6], as well as its multifractionality [7].

Spatial graded composites have a wide range of uses. The matrix of such composites includes two or more components. The relative volume fractions of the components or the sizes of the filler particles change in the volume of the matrix according to a fixed law [8].

Thermal anisotropy of composites with spatial gradation leads to the appearance of effective thermal barriers and nonlinear heat fluxes.

The numerical experiment method is practically the only method for predicting the thermophysical properties of both composite materials with a randomly inhomogeneous spatial distribution of complex-shaped filler particles and spatial-gradient materials.

A necessary condition for the possibility of calculating temperature profiles in heterogeneous systems of complex geometry, along with using the concept of effective thermal conductivity, is the numerical calculation of the distribution of local heat fluxes (LHF).

Our work is devoted to the construction of maps of local heat fluxes of two-component composites both for single inclusions and for an array of inclusions with variable distribution characteristics in the matrix. In this work, we will investigate such a characteristic of the direction of local heat fluxes as the dependence of the weighted average angle of deviation on the relative concentration of inclusions. The study of the integral characteristics of two-component composite in this work was limited to the analysis of approximation formulas for the dependence of the effective thermal conductivity on the relative concentration.

2. Materials and methods

In this work, for numerical experiments, we used a quasi-two-dimensional model of a composite material (CM), i.e. a model in which heat transport was carried out only along two coordinates, and there was no heat transfer along the third coordinate. For this model, we numerically solved the Fourier differential heat conduction equation with boundary conditions of the first kind, namely, with fixed temperatures at the outer boundaries of the computational domain. In the absence of heat sources or sinks in the CM volume, the Fourier equation has the form

$$\nabla[\lambda(x, y), \nabla T(x, y)] = 0, \quad (1)$$

where $T(x, y)$ is the temperature field and $\lambda(x, y)$ is the thermal conductivity. The value $\lambda(x, y)$, depending on the x and y coordinates, can be equal to the thermal conductivity of the matrix material (λ_m) or the thermal conductivity of the filler material (λ_f).

The model used a square shape for both filler particles (inclusions) and the matrix, which made it possible to use the finite difference method on an orthogonal grid with a step h in both coordinates to solve equation (1). The features of the calculation algorithm are described in detail in our previous works [9, 10]. As a result of solving equation (1), the temperature field and the field of vectors of the heat flux density were calculated according to the following equation

$$f(x, y) = -\lambda(x, y)\nabla T(x, y) \quad (2)$$

In the course of numerical experiments, we simulated local heat fluxes both with the placement of heat-conducting inclusions in the matrix, and with exactly the same placement of heat-insulating inclusions in the matrix, and compared the simulation results in these two cases. The ratio of thermal conductivity of CM components λ_f / λ_m was chosen for the case of heat-conducting inclusions approximately equal to the ratio of thermal conductivity of reinforcing steel and cement-sand stone (50 : 1), and for the case of heat-insulating inclusions equal to the ratio of thermal conductivity of expanded polystyrene and cement-sand stone (1 : 21).

The distribution of inclusions in the matrix was random and equiprobable. Both the matrix and the inclusions were square. The algorithm for creating the CM model made it possible to set the size of the matrix equal to a', the size of the inclusion equal to b' and the minimum allowable distance between the inclusions equal to d'. The indicated dimensions were expressed in whole numbers of steps of the computational grid h. The concentration of inclusions with was determined as the ratio of the total area occupied by the inclusions to the area of the matrix $c = N_f (b'/a')^2$, where N_f is the number of inclusions that was specified in the simulation. In CM model, unless otherwise specified, the side of the square of the matrix was chosen to be $128h$. When analyzing the results of numerical experiments, we operated on the following quantities: the relative size of the inclusions $b = b'/a'$ and the relative minimum distance $d = d'/b'$.

3. Results and discussion

As we have shown in [10], visualization of the density and LHF directions in CM models helps to correctly interpret the results of numerical experiments. Figures 1 – 4 show typical fragments of the calculated LHF maps in CM model with heat-insulating and heat-conducting inclusions. The degree of blackening of the map background is inversely proportional to the density of the LHF vector. Arrows show the direction of the vectors. In the area of the lowest-intensity LHF, to contrast with the dark background, the color of the arrows was changed to white. The temperature difference on all maps was applied from left to right.

Figure 1 shows the LHF map in the vicinity of a solitary heat-insulating inclusion of size $b = 0.125$. Figure 2 presents the LHF map in the vicinity of a solitary heat-conducting inclusion of equal size.

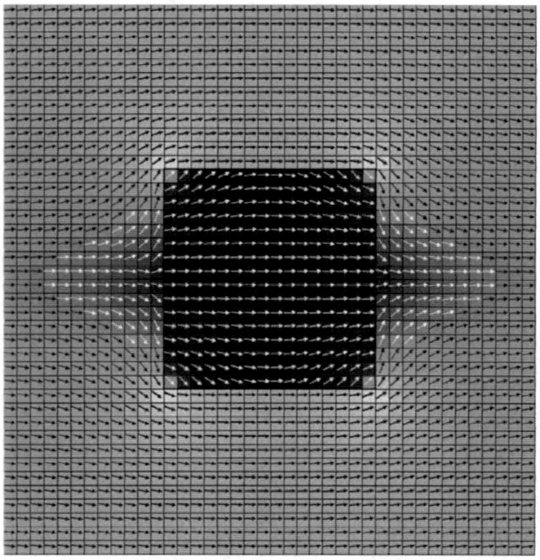

Figure 1. LHF map in the vicinity of heat-insulating inclusion.

Figure 2. LHF map in the vicinity of solitary heat-conducting inclusion.

Figure 1 shows that LHF pass around the heat-insulating inclusion, forming regions with a low flux density (the so-called dark matrix) in the front and rear parts of the matrix, and regions with increased flux density along the upper and lower sides of the inclusion. In the case of a heat-conducting inclusion, as shown in figure 2, LHF from nearby regions of the matrix are concentrated inside the inclusion, with a dark matrix being formed along the upper and lower sides of the inclusion. The distortion of the LHF direction in the vicinity of inclusion in both cases is recorded at a distance of the order of its size. Thus, if two inclusions of size b' are at a distance exceeding $2b'$, they can be considered as non-interacting. The maximum concentration of non-interacting inclusions in the two-dimensional case with a uniform distribution of inclusions over the matrix will be $1/9 \approx 0.11$. The nearest environment of each inclusion for the case of higher concentrations will affect the density and direction of the LHF in its vicinity.

Figure 3 shows a fragment of the LHF map in a matrix with interacting heat-insulating inclusions of size $b = 0.078$. Figure 4 presents a fragment of the LHF map in a matrix with exactly the same arrangement of heat-conducting inclusions.

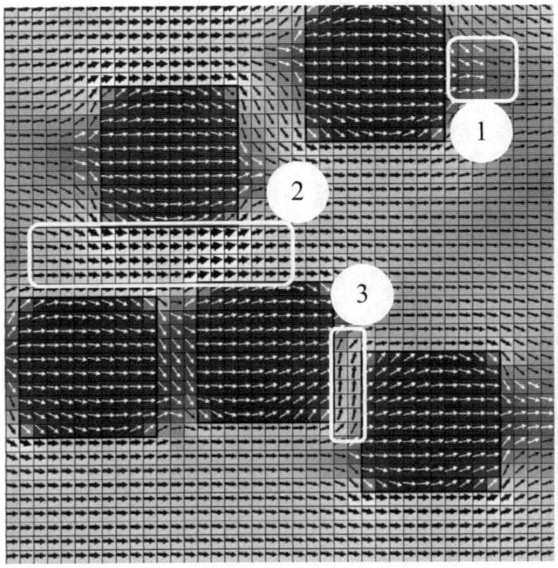

Figure 3. Fragment of the LHF map in the model with heat-insulating inclusions. Legend: 1 – dark matrix area; 2, 3 – regions of induced heat-conducting channels. Gray scale of LHF density: min ▬ max.

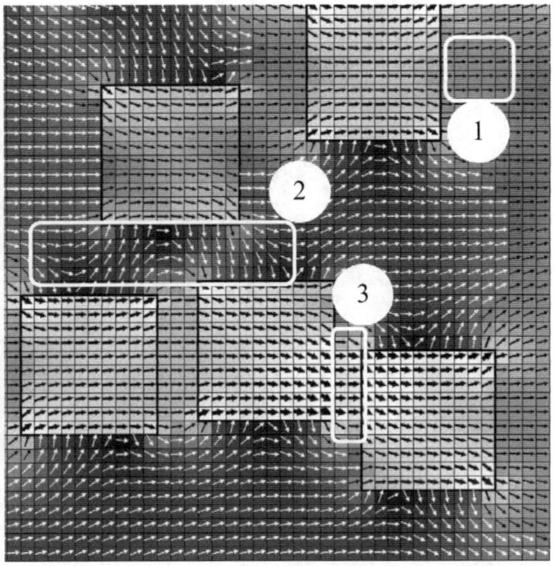

Figure 4. Fragment of the LHF map in the model with heat-conducting inclusions. The same areas are highlighted as in figure 3. Legend: 1 – unperturbed matrix area; 2 – dark matrix area; 3 – region of the induced heat-conducting channels.

Analysis of figure 3 shows that local heat fluxes tend to bend around heat-insulating inclusions and move along the matrix between them. The interaction of heat-insulating inclusions, as we have shown in [12], leads to the formation of extended induced heat-conducting channels, i.e. fragments of the matrix, through which the bulk of heat transfer is carried out. These channels have a sinuous shape, i.e. their length exceeds the distance between the hot and cold faces of the sample, and the local heat flux in these channels is directed at a certain angle with respect to the direction from the hot to cold faces.

In the case of heat-conducting inclusions (figure 4), local heat fluxes tend to concentrate in inclusions, and they overcome the heat-insulating matrix with a sufficient concentration of inclusions, migrating from inclusion to inclusion. Thus, also in this case, the local heat flux turns out to be directed at a certain angle with respect to the direction from the hot to the cold face.

Figures 5 and 6 show an example of the distribution of the local heat fluxes density with the same placement of inclusions of size $b = 0.039$ and concentration $c = 0.244$ with a minimum distance $d = 0.2$. The light color indicates the CM areas along which the bulk of heat transfer occurs (for a given set of CM parameters, this fraction is about 70%).

Figure 5 shows that in this case, the induced heat-conducting channels are formed between parallel chains of heat-insulating inclusions, while in Figure 6 it can be seen that the induced heat-conducting channels are formed by chains of closely spaced heat-conducting inclusions. In both cases, the direction of the induced heat-conducting channels does not always coincide with the direction from the hot to the cold side; therefore, the path of the heat flow from the left to the right side will be greater than the distance between the sides, which leads to an effective increase in the thermal resistance of the material.

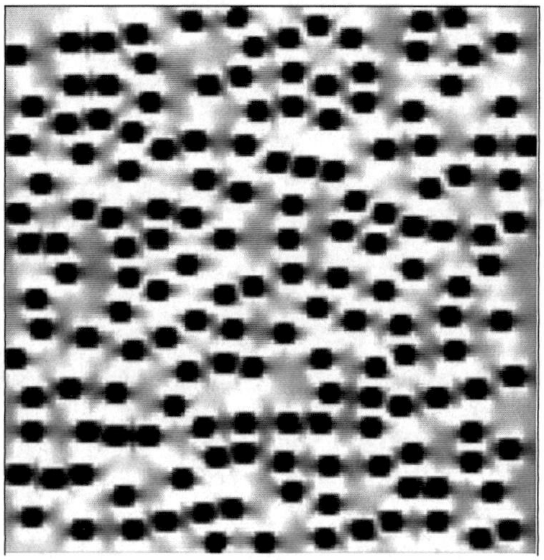

Figure 5. Density map of LHF in CM with heat-insulating inclusions.

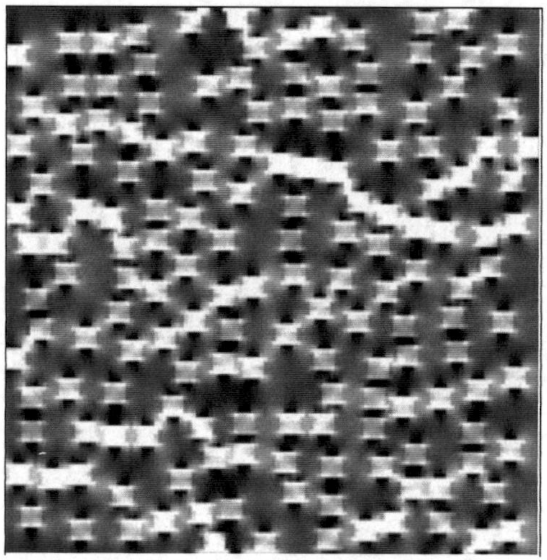

Figure 6. Density map of LHF in CM with heat-conducting inclusions.

Gray scale of LHF density: min max.

In order to describe the effective increase in the thermal resistance of the material, we introduced in [10] the value of the weighted average angle (WAA) of deviation of the direction of the local heat flux from the direction of the temperature difference on the sample faces

$$<\varphi_W>=\left(\sum_{i,j}\varphi_{i,j}\left|f_{i,j}\right|\right)\cdot\left(\sum_{i,j}\left|f_{i,j}\right|\right)^{-1}, \tag{3}$$

where $|\boldsymbol{f}_{i,j}|$ is the modulus of the LHF density vector, calculated at the node of the computational grid with numbers i and j; $\varphi_{i,j}$ is the angle between the direction of the LHF vector and the direction of the temperature drop on the faces of the samples at the same node of the computational grid.

Numerical simulation allowed us to confirm that the weighted average angle $<\varphi_W>$ has a significant correlation with the effective thermal conductivity of the sample [10].

Figure 7 shows the dependence of $<\varphi_W>$ on the concentration of the filler for heat-insulating and heat-conducting inclusions of various sizes. The placement of heat-conducting inclusions in the CM leads to an increase in the angles of LHF deviation from the direction of the temperature drop at the edges of the samples as compared to the placement of heat-insulating inclusions. The monotonic growth of $<\varphi_W>(c)$ in the case of a heat-insulating filler is due to the fact that with an increase in the filler concentration, LHF are forced to bend around more obstacles. In this case, the path of the LHF increases. This should lead to a sublinear decrease in the effective thermal conductivity of CM λ_e. For a heat-conducting filler, the $<\varphi_W>(c)$ dependence has a sharp maximum, which shifts to the right with an increase in the size of the inclusions. In this case, LHF migrate from inclusion to inclusion, and $<\varphi_W>$ reflects the probable relative position of the inclusions. An initial growth in $<\varphi_W>$ with an increase in filler concentration is due to the fact that the number of inclusions available for migration of LHF, located at significant angles to the direction of the temperature drop, increases. Because in this case $\lambda_f / \lambda_m >> 1$, then even with the lengthening of the total path of LHF, the thermal resistance along this path decreases. When a certain concentration is reached, the probability of finding an inclusion located in the direction of the temperature difference increases, and as a result $<\varphi_W>$ decreases. Thus, it is natural to expect super linear growth of λ_e.

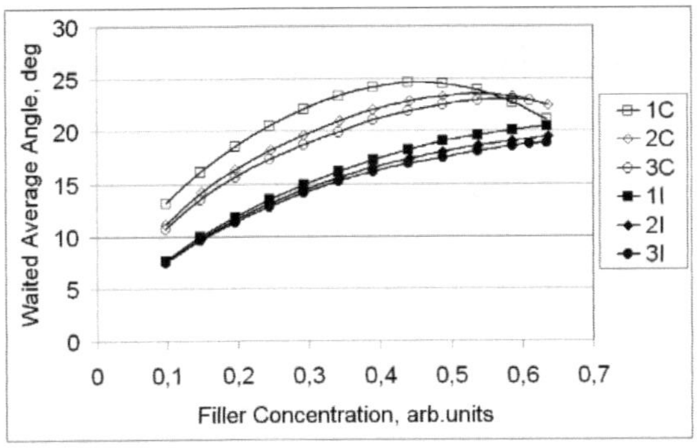

Figure 7. Dependences of $<\varphi_W>$ on the filler concentration. Filler dimensions: $1 - b = 0.039$; $2 - b = 0.0625$; $3 - b = 0.078$. Symbol «**C**» – heat-conducting inclusions; Symbol «**I**» – heat-insulating inclusions. For all dependencies: $d = 0$.

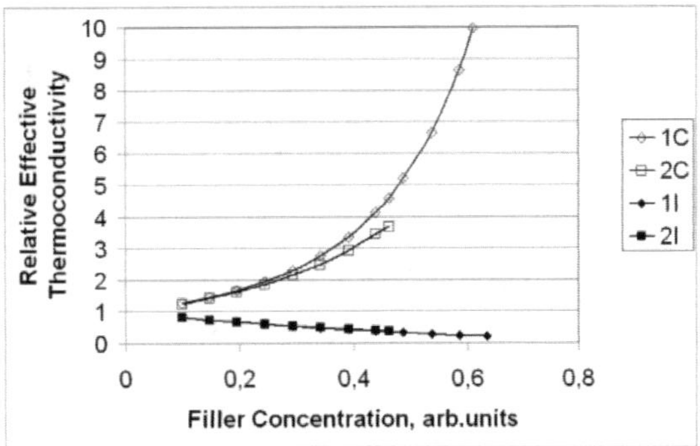

Figure 8. Dependences of λ_e / λ_m on the filler concentration. Minimum distance between filler particles: $1 - d = 0$; $2 - d = 0.2$. Symbol «**C**» – heat-conducting inclusions; Symbol «**I**» – heat-insulating inclusions. For all dependencies: $b = 0.078$.

Figure 8 shows the dependences of the relative effective thermal conductivity of CM λ_e / λ_m on the filler concentration for heat-insulating and heat-conducting inclusions with a size of $b = 0.078$ at two minimum distances d. In previous works [9, 10] we proposed an approximation of the dependence λ_e / λ_m on c for the case of a heat-insulating filler in the form

$$\frac{\lambda_e}{\lambda_m} = 1 - \frac{\alpha c}{\sqrt{(1+\beta c)^3}}, \tag{4}$$

where α and β are fitting parameters. In the case of a heat-conducting filler, approximation (4) turns out to be inapplicable. Taking into account our new results, we propose to consider the approximation for the dependences of λ_e / λ_m on c in the form

$$\frac{\lambda_e}{\lambda_m} = \alpha + (1-\alpha)^{(1-c)}\left[\beta\left(\frac{\lambda_f}{\lambda_m} - \alpha\right)\right]^c, \tag{5}$$

where α and β are also fitting parameters. In this case, for the heat-insulating filler ($\lambda_f / \lambda_m << 1$) the sign of the parameter α will be negative, and for the heat-conducting filler ($\lambda_f / \lambda_m >> 1$) the sign of the parameter α will be positive. The approximation parameters α, β and the coefficient of determination R^2 for the curves in figure 8 are given in table 1.

Table 1. Fit parameters and goodness of fit.

	Heat-insulating inclusions		Heat-conducting inclusions	
	$b = 0.078$; $d = 0$	$b = 0.078$; $d = 0.2$	$b = 0.078$; $d = 0$	$b = 0.078$; $d = 0.2$
α	-0.3000	-0.0111	0.7088	0,4284
β	0.8412	2.247	1.671	0.5007
R^2	0.9999	0.9996	0.9993	0.9998

4. Conclusions

1. With significant differences in the thermal conductivity of inclusions located in the composite matrix, and the thermal conductivity of the matrix itself, induced heat-conducting channels are formed in the composite material. The major part of heat transfer through the composite occurs through these channels. The ratio of the thermal conductivity of the inclusions and the matrix decisively affects the location of the induced heat-conducting channels and the direction of local heat fluxes. 2. The weighted average angle of deviation of the heat flux from the direction of the temperature difference between the faces of the samples is considered as an integral characteristic of the local heat fluxes direction. The explanation of the peculiarities of the dependences $<\varphi_W>$ on the filler concentration can be carried out using probabilistic ideas about the position of neighboring inclusions. 3. We have obtained a general formula for approximating the dependence of the effective thermal conductivity of the composite on the filler concentration for the cases of heat-insulating and heat-conducting inclusions. 4. The dependences $<\varphi_W>$ and λ_e / λ_m on the filler concentration are due to the peculiarities of heat transfer in composites at the micro level. The results obtained in this work can be used both for modeling the properties of composite materials and for optimizing technological processes for the manufacture of composites with specified thermophysical properties.

References

[1] Agraval A, Satapathy A 2015 Mathematical model for evaluating effective thermal conductivity of polymer composites with hybrid fillers *Int. J. Therm. Sci* **89** 203-209

[2] Angle J-P, Wang Z, Dames C and Mecartney M-L 2013 Comparison of two-phase thermal conductivity models with experiments on dilute ceramic composites *J. Am. Ceram. Soc* **96(9)** 2935-2942

[3] Cao C, Yu A and Qin Q-H 2012 Evaluation of Effective Thermal Conductivity of Fiber-Reinforced Composites *Int. J. Archit. Eng. Constr* **1(1)** 14-29

[4] Khan K-A, Khan S-Z and Khan M-A 2016 Effective thermal conductivity of two-phase composites containing highly conductive inclusions. *J. Reinf. Plast. Compos* **35(21)** 1586-1599

[5] Ngo I-L, Byon C 2015 A generalized correlation for predicting the thermal conductivity of composite materials *Int. J. Heat Mass Tran* **83** 408-415

[6] Duan H-L, Karihaloo B-L, Wang J and Yi X 2006 Effective conductivities of heterogeneous media containing multiple inclusions with various spatial distributions. *Phys. Rev B* **73(17)** 174203-13

[7] Böhm H-J, Nogales S 2008 Mori–Tanaka models for the thermal conductivity of composites with interfacial resistance and particle size distributions *Compos. Sci. Technol* **68(5)** 1181-1187

[8] Yin H-M, Paulino G-H, Buttlar W-G and Sun L-Z 2005 Effective thermal conductivity of two-phase functionally graded particulate composites. *J. Appl. Phys* **98(6)** 063704

[9] Zaginaylo I, Maksimeniuk Ya and Pysarenko A 2017 Two-dimensional numerical simulation study of the effective thermal conductivity statistics for binary composite materials *Int. J. Heat Technol* **35(2)** 364-370

[10] Pysarenko A and Zaginaylo I 2019 Numerical Simulation of the Heat Conductivity of Randomly Inhomogeneous Two-Dimensional Composite Material *Hauppage, New York: Nova Science Publisher's, Inc.* 176

[11] Zaginaylo I V, Maximenouk Ya A and Pisarenko A N 2016 Induced thermally conductive channel influence on the formation of the insulating properties of binary composite materials *Construction, material science, mechanical engineering* ed V I Bolshakov (Dnipro) **92** 56-61

Analysis of the structure formation processes of building composites by geometric methods

S Semenova[1], A Kolesnikov[1], G Kyrylenko[1], T Oliynik[1]

[1] Department of Chemistry and Ecology, Odessa State Academy of Civil Engineering and Architecture, Didrikhsona st., 4, 65029, Odessa, Ukraine

E-mail: semenovablacksea@gmail.com

Abstract. The work considers a method of structure formation modeling in composite materials based on geometric methods. For this purpose, a differential system is written for the fractions of the volume, surface, linear and point phase of the structure forming components of the material. To study it, a decomposition into a gradient and generalized-conservative component was used. Approaches to modeling of the processes of structure formation and destruction based on the theory of gradient systems, which take into account the structural potentials for the surface and volumetric components of the composite material, were proposed. The hypothesis according to which the formation of a high-strength structure of the material is associated with the implementation of the oscillatory mode for the indicator of the fraction of the volumetric phase was proposed. The mechanism of increasing the strength of the composite binder dough based on the selective preservation of the strongest intercrystalline contacts was discussed.

One of the main features of the processes of structure formation in dispersed systems and, in particular, in the binder dough of a composite material, is the appearance of new physicochemical objects - internal interfaces. Together with other structural elements - clusters of binder, filler and aggregate, they form an integral structure of the composite material [1]. All of these objects are geometrically complex, often fractal-like. One of the demonstrations of this complexity is a significant difference in the shape of the formed crystals of the new phase from the spherical one. The shape of the polycrystalline phase, like the crystals themselves, is characterized by the presence of edges and apexes, i.e. is a three-dimensional body with a certain volume, interface area, rib frame length, number of apexes (Fig. 1).

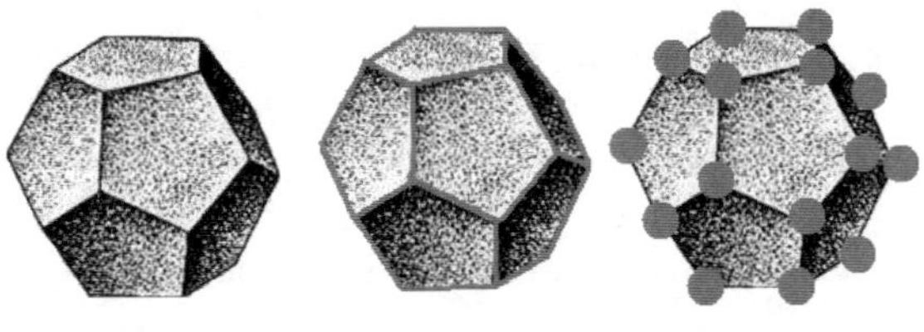

Fig. 1. Schematic representation of geometric features of the emerging new phase: S – surface area, L – linear skeleton, P – apex points.

The presence of geometric features affects the physical processes of the growth of crystals of the new phase, and their physicochemical characteristics. The research of structure parameters by geometric method, using the representation of the polycrystalline phase as fractal objects [2], can be implemented both in composites with a crystallization mechanism of structure formation (gypsum binders) and at a different scale level, for microcrystalline formations prevailing in cement binder dough.

To display the features of the structure formation of objects of complex (fractal-like) geometry, we consider the vector field specified by the coordinates $(y_0(t), y_1(t), y_2(t), y_3(t_0))$. Each of its components corresponds to the proportion of point, linear, surface and volumetric components of the dispersed system [3]. Hereinafter, in order to consider the dynamic processes of structure formation and destruction, the possibility of changing these values with time as a result of physicochemical processes was taken into account. In the case of mineral binders, hydration and crystallization processes will be the main ones. The corresponding n-dimensional vector field $(y_0(t), y_1(t), \ldots , y_n(t))$ is quite definite if the differential system (1) is given:

$$\frac{dy_i}{dt} = f_i(y_0,\ldots, y_3), \quad i = 0,\ldots,n, \tag{1}$$

and the corresponding initial conditions $y(t_0) = y_i^0$. System (1) can be considered in the general case together with balance and stoichiometric equations. All values y_i are nonnegative and bounded from above by one; therefore, in the correctly constructed model (1) according to the Bohl-Brauer theorem, the system has at least one fixed point - the equilibrium state [4]. When studying system (1), it is useful to single out the gradient, corresponding to the term $-grad\,V(y)$, and the generalized conservative component $Q(y)$ (2):

$$\frac{dy}{dt} = -grad\,V(y) + Q(y). \tag{2}$$

Let condition (3) be satisfied:

$$Q \cdot grad\,V(y) = 0. \tag{3}$$

Generalized conservative component $Q(y)$ associated with motion along the lines of the potential level $V(y)$. Since from (1) and (2) it follows (4):

$$Q(y) = f(y) + grad\,V(y), \tag{4}$$

then for $V(y)$ we find a partial differential equation (5) similar to the Hamilton-Jacobi equation:

$$(f(y) + grad\,V(y)) \cdot grad\,V(y) = 0. \tag{5}$$

In equation (5), the transition to vector designations was carried out. Equations (4) and (5) allow one to carry out expansion (2).

The system described using (1) has a rather general form: in addition to asymptotically stable (physically realizable) equilibria, unstable, as well as limit cycles, and strange attractors can exist in it.

The primary model of structure formation in composite materials can be built on the basis of the gradient representation (6)

$$\frac{dy_i}{dt} = -\frac{\partial V(y,c)}{\partial y_i}. \tag{6}$$

Equation (6) includes the potential $V(y_2,y_3)$ for volumetric (y_3) and surface (y_2) components of the material, which is sufficient to consider many problems of structure formation. Considering for this case the expansion in a Taylor series, and carrying out a number of transformations associated with the diagonalization of the matrix of coefficients [5,6], we obtain as a result the equation of an umbilical surface, for example, hyperbolic (7):

$$V(y_2, y_3) = y_2^3 + y_3^3 + wy_2x_3 - uy_2 - vy_3, \tag{7}$$

where w, u, v – control parameters that depend, for example, on the conditions of structure formation.

In some cases, for example, when discussing the problems of material destruction, it is sufficient to take into account one geometric parameter that serves as an order parameter for the system under consideration - the fraction of surface states y_2. Here the calculation is extremely simplified and the analogy with the theory of phase transitions is strengthened. Thus, using potential (8):

$$V(q_2) = \frac{q^4}{4} + \frac{uq^2}{2} + vq, \quad (8)$$

corresponding to an assembly catastrophe, it is possible to cover, at least on a qualitative level, such effects as the destruction of a material that is initially in a structurally metastable state (Fig. 2), and many similar effects.

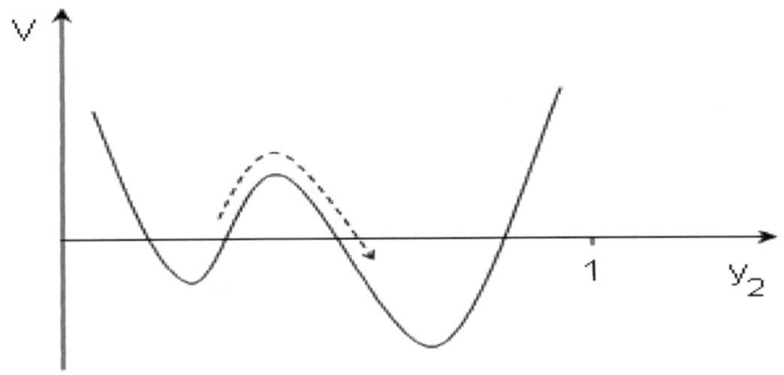

Fig. 2. Qualitative display of the material destruction process: the fraction of the surface component y_2 increases.

One of the possible models of material destruction obtained using the above approach is a sharp increase in the proportion of surface, linear and point components: y_2, y_1 and y_0 respectively.

If in a real system there is no energy supply, then one should expect the approximation and realization of some stable stationary regime, which happens during structure formation in material media. When energy is supplied from the outside, structural rearrangements can have a cyclic character, which corresponds to asymptotically stable limit cycles in system (1) (Fig.3, C). The role of such an energy source can be played by physicochemical processes at the interface, for example, an increase in temperature during wetting, the phenomena of hydration and dissolution. In view of the considered decomposition procedure (2) of system (1), of interest is the Rebinder - Segalova theory associated with the interpretation of the slowdown in strength growth during hardening of gypsum binders, as well as the sometimes observed (especially for lime binders) appearance of secondary minima on the plastic strength plots. The processes of structure formation and destruction are presented in this theory as a single whole, with a sequential alternation of phases of the prevailing growth processes of a new phase: fouling of the primary framework of the coagulation-crystallization structure, the occurrence of internal stresses, partial destruction of the primary framework, secondary enhancement of the growth of neoplasms. Thus, near the "Rebinder point" corresponding to the minimum of the potential $V(y)$, corresponding to the structural equilibrium of the composite binder, the generalized conservative component $Q(y)$, corresponding to rotational motion around the equilibrium state, does not vanish (Fig. 3, A,B).

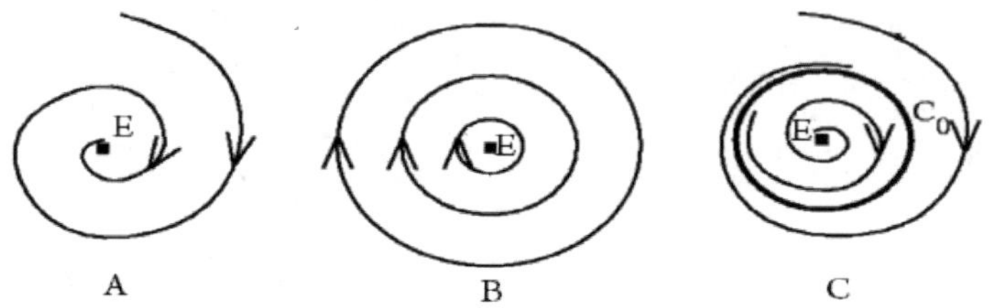

Fig. 3. Elements of the phase portrait of system (1): equilibrium states E (A-stable focus, B-center, C-realization of an unstable focus and limit cycle).

The generalization of the Rebinder-Segalova theory [7], taking into account the considered concepts, allows us to put forward a geometric hypothesis about the formation of binding properties, which are emerged both at the level of individual mineral binders and in composite mixtures. Next consider the ratio of the proportions of the volumetric and surface components of the binder in an individual form or in the compound of a composite material (9):

$$y_{32} = \frac{y_3}{y_2}. \tag{9}$$

The results of Rebinder's work considered above, the research of the hardening of lime binders indicate the oscillatory nature of the change in this value during the hardening period (Fig. 2). The proposed hypothesis is as follows: the formation of a stone-like body and the hardening of a composite binder is closely related to the processes of cyclic changes of the parameter. y_{32}. Phase changes are associated with the dissolution of the original binder, the formation and growth of a new phase (decrease y_{32}), the formation of a primary coagulation-crystallization framework and the gradual accretion of gypsum binder crystals, overgrowth of the primary framework (increase y_{32}), its partial destruction (decrease y_{32}), the formation of a new framework (increase y_{32}) [8]. The stabilization of the disperse system corresponding to the hardened composite is thus carried out as a result of a damped oscillatory process (Fig. 4, 3), and the corresponding phase plane ($y_{32}, \dot{y}_{32}$) is characterized by the presence of a stable focus.

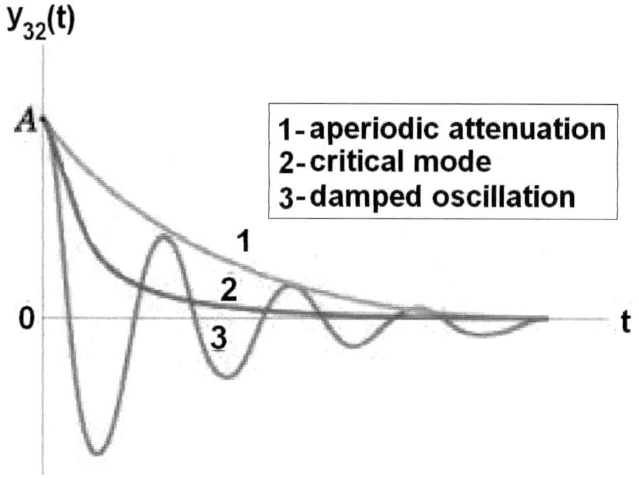

Fig. 4. The nature of temporary changes y_{32} during hardening.

According to this hypothesis, the processes corresponding to the trajectories (Fig. 4, 1 and 2) lead to the formation of a material of low strength. For the formation of a material of increased strength, it is necessary to carry out an oscillatory process for y_{32}. One of the interpretations of the role of oscillations in the formation of strength can be based on the theory of the selection process [9,10]. The destruction of the formed primary crystallization framework occurs due to the destruction of mechanically unstable, weakened bonds between the growing crystals of the gypsum composite, which in the next phase of the oscillatory process are replaced by new ones.

The proposed hypothesis is confirmed, in particular, by the study of microscopic preparations of hardening gypsum G-5 as a model system (Fig. 5).

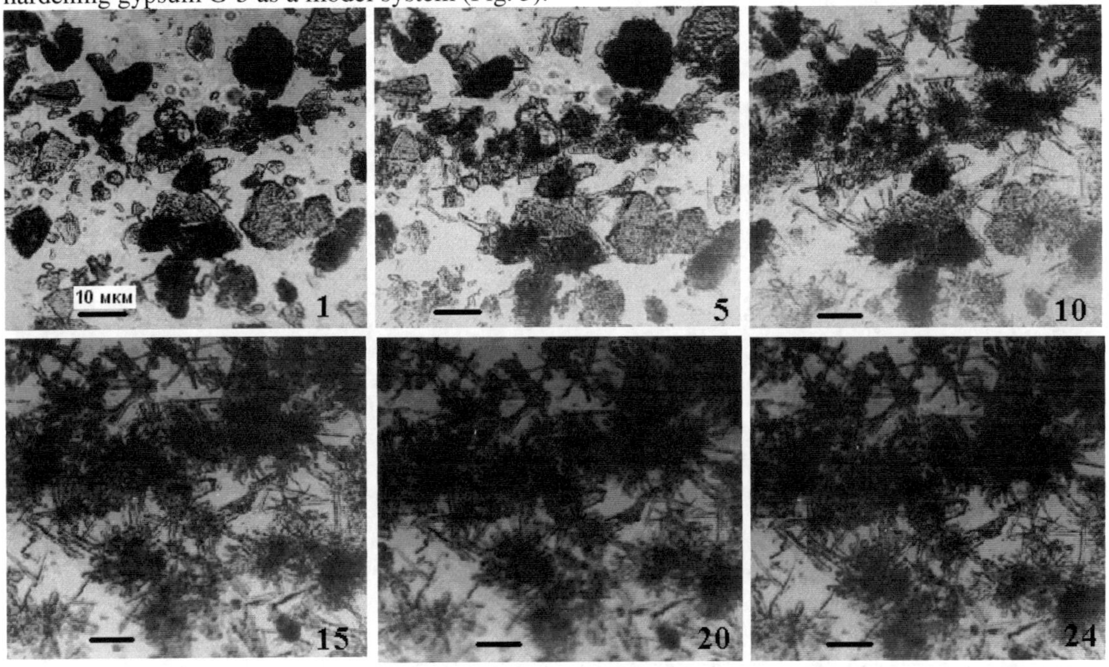

Fig. 5. Images of hardening gypsum preparations. The figures show the time from mixing (min.) of the binder dough and the scale.

By the method of computer image processing, the values of the volumetric component y_3 and surface component y_2 are obtained, and the corresponding ratio y_{32} is determined. Its change in the process of binder hardening is reflected in the graph (Fig. 6).

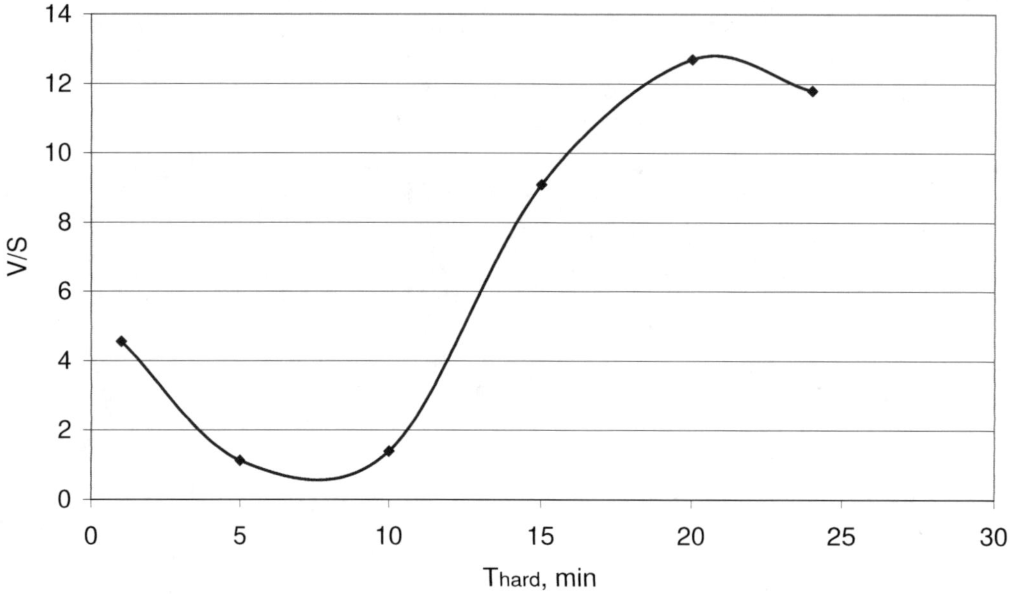

Fig. 6. Variations in the parameter $y_{32} = V/S$ during hardening.

Figure 2 shows the initial phases of the oscillatory process. Figure 2 shows the initial phases of the oscillatory process. Since a high-density crystallization structure is formed during the hardening of the material (Fig. 1, 24 min.), the possibility of optical methods is significantly limited. It should also be noted that there is a significant difference in the geometric properties of the quasi-two-dimensional (micropreparation) and three-dimensional crystallization structures. For further registration of the oscillatory process, indirect measurements of the desired parameter $y_{32} = V/S$, are applicable, for example, based on the electrical characteristics of the binder dough and plastic strength.

Thus, consideration of the geometric properties of materials from a quantitative and dynamic point of view seems to be one of the ways to analyze the processes of structure formation of binders. On the basis of the proposed method of modeling structure formation, a hypothesis is considered, according to which the oscillatory modes of hardening of the composite binder dough are associated with the selection of bonds, which is accompanied by the formation of high-strength structures of the composite material.

Methods of geometric dynamics, in our opinion, allow not only to investigation the peculiarities of the structure formation of binders, but also to pass to adequate problems of control and optimization of the structure formation of building composites.

Reference

[1] Solomatov V, Vyrovoy V, Dorofeyev V, Sirenko A 1991 Composite materials of reduced material consumption *Kiev: Budivelnyk*

[2] Feder E 1988 Fractals *New York: Springer Sience+Business Media*

[3] Dovgan I, Kolesnikov A, Semenova S, Kirilenko G 2011 Description of the geometric properties of dispersed systems using a polynomial *Bulletin of OSACEA* Issue **43** Odessa:Vneshreklamservice pp 115-119

[4] Pokrovsky I 1970 Lectures on the theory of ordinary differential equations *Moscow: Nauka*

[5] Haken G 1984 Synergetics *Berlin: Springer-Verlag*

[6] Gilmore R 1993 Catastrophe Theory for Scientists and Engineers *Dover Publications*

[7] Segalova E, Rebinder P 1962 The appearance of crystallization structures of hardening and conditions for the development of their strength *Moscow: Stroyizdat*
[8] Kolesnikov V 1990 Structure formation in dispersed systems *Odessa: OCEI*
[9] Eigen M, Schuster P 1979 Hypercycle A Principle of Natural Self-Organization *Berlin: Springer-Verlag*
[10] Kuznetsova V, Rakov M 1987 Self-organization in technical systems *Kiev: Naukova dumka*

Increasing the water resistance of the composite gypsum binder for arbolito concrete due to hydrofobization

O S Shynkevich[1], D S Linnik[2,5], I V Zaginaylo[3] and G G Bondarenko[4]
[1] Department of processes and machine in the technology of building materials, Building and Technology Institute, Odessa State Academy of Civil Engineering and Architecture, Odessa, Ukraine
[2] Department of metal, wood and plastic structures, Institute of Civil Engineering, Odessa State Academy of Civil Engineering and Architecture, Odessa, Ukraine
[3] Department of Physics, Institute of Civil Engineering, Odessa State Academy of Civil Engineering and Architecture, Odessa, Ukraine
[4] Nikolaev Construction College KNUSA, Nikolaev, Ukraine
[5] fgg109m@gmail.com

Abstract. The article discusses the problem of increasing the water resistance and strength of a composite gypsum binder for concrete. The effect of complex modification of a composite gypsum binder with a micro-reinforcing, polymineral micropuzzolanic additive filler and superplasticizer from Sika was investigated. The use of filler additives in optimal ratios made it possible to obtain a composite gypsum binder with a softening coefficient of 0.9.

1. Introduction

Energy efficiency, resource saving, lower operating costs and increased housing comfort is a priority in construction practice. Development trends of the modern construction industry are focused on increasing competitiveness, developing and introducing new technological solutions that provide resource and energy saving, high technical, economic and consumer indicators of product quality. The optimization of product quality indicators is ensured through the use of durable, sound, high-quality building materials. An important requirement is the environmental friendliness of building products, which can be achieved through the use of appropriate materials and compliance with the closed-cycle technological mode [1].

The materials of this type relates gypsum concrete - concrete based composite gypsum binder [2-4]. Gypsum materials and products are advanced building materials due to the simplicity and low energy consumption of production. For the production of 1 ton of gypsum binder, 4.5 and 4.9 times less fuel and electricity are used than for the production of 1 ton of Portland cement. The hardening time of a gypsum binder is 20–25 times less than that of Portland cement; a set of design strength for products based on gypsum binder is 30–40 times faster than that based on Portland cement. In addition, materials containing gypsum are high-tech, because already 15-20 minutes after the product is formed, it can be removal of formwork, because of this the mould reusability increases in 8-10 times, and manufacturing techniques are less energy-consuming in contrast to the use of other types binders, including Portland cement.

Ukraine has sufficient reserves of natural gypsum raw materials and a huge amount of gypsum-containing waste. One of the disadvantages of gypsum binders is the low water resistance. One of the modern effective ways to increase water resistance can be a modification of the structure of the binder with polymineral microdispersed filler additives [5-7].

The use of new types of volumetric hydrophobizators will provide a further improvement in quality. The compositions of hydrophobizators are available in the form of concentrates or ready-to-use solutions in aqueous or organic solvents. They create a water-repellent effect, while the vapor permeability of such composites is practically not reduced. The introduction of water repellents does not change the color of the product, hydrophobization significantly increases the frost resistance of any material. These properties allow the use of water repellents for composite gypsum binder.

Objective: to development of a complex of highly dispersed polymineral additives, a micro-reinforcing filler additive and a superplasticizer for a composite gypsum binder with improved properties.

Research Objectives:

-selection and justification of the components of a complex polymineral micropuzzolanic additive and micro-reinforcing filler additives for composite gypsum binder;

- selection of the ratio of the finely divided components of the complex polymineral micropuzzolanic additive and the micro-reinforcing filler additive for the composite gypsum binder using experimental statistical modeling methods based on the mathematical theory of experimental design;

- evaluation of the interdependent on water resistance and compressive strength of the composite gypsum binder complex of a polyfunctional modifier, and a volume hydrophobization.

2. Materials and research methods

The following components were used as binder components: building gypsum G5 produced by PJSC Gypsum, cement pozzolanic non-additive produced by OJSC Euro Cement Ukraine PC-I-500D0 according to DSTU B B.2.7-46: 2010.

To obtain a composite gypsum binder, highly active metakaolin TU U 14.2-36363275-001: 2009 manufactured by LLC Meta-D, silica fume produced by Elkem AS EN 13263-1 as active pozzolanic additives were used.

Wollastonite TU 5777-006-40705684-2003 of different dispersion produced by Geocom CJSC was used as a micro-reinforcing additive.

Sika ViscoCrete 225 superplasticizer proshka VP SIA 162 (1989) and prEN 934-2, manufactured by Sika trademark, was used as an additive of retarder.

Composite gypsum binders' samples were tested for strength at the age of 28 days according to DSTU B B.2.7-187: 2009, followed by drying to constant weight. The determination of the softening coefficient was carried out according to TU 21-0284757-90.

3. Planning a Six-Factor Experiment

The choice of the experimental plan for solving specific materials science problems is dictated by the specific conditions of these problems and the requirements for the results of their solution. For the purpose of work, a 24-point plan in the form of "triangles on a cube" was applied. The use of MTQ plans is subject to the following provisions. Used to obtain modified composite gypsum binders, additives modifiers and fillers are mixtures with q various substances of different microdispersion.

The composition of such mixtures can be specified by the concentrations of the components in the form of mass, volume or molar fractions (percent) v_i. Systems whose properties are determined by the group of dependent mixed factors $\mathbf{v} = (v_1 \ldots v_q)$ and the group of independent technological factors $\mathbf{x} = (x_1 \ldots x_k)$, are called systems "composition, mixture - properties" or "composition - technology - properties" and are designated by the abbreviation MTQ. To describe MTQ systems, an experimental statistical (ES) model is used, which has a general view:

$$Y = \underbrace{\begin{array}{|c|}\hline A_1v_1+A_{12}v_1v_2 \\ A_2v_2+A_{13}v_1v_3 \\ A_3v_3+A_{23}v_2v_3 \\ \hline\end{array}}_{(a)} + \underbrace{\begin{array}{|c|}\hline D_{14}v_1x_4+D_{15}v_1x_5+D_{16}v_1x_6 \\ D_{24}v_2x_4+D_{25}v_2x_5+D_{26}v_2x_6 \\ D_{34}v_3x_4+D_{35}v_3x_5+D_{36}v_3x_6 \\ \hline\end{array}}_{(b)} + \underbrace{\begin{array}{|c|}\hline b_{44}x_4^2+b_{45}x_4x_5 \\ b_{55}x_5^2+b_{46}x_4x_6 \\ b_{66}x_6^2+b_{56}x_5x_6 \\ \hline\end{array}}_{(c)} \quad (1)$$

ES models are calculated according to the results of experiments using a standard version of PC COMPEX program, which implements a sequential regression analysis with a generating experimental error. ES models can be used for decision-making only after the algebraic calculation of coefficient estimates is supplemented by statistical analysis of both individual coefficients and the model as a whole. Statistical control makes it possible with a given degree of probability to evaluate the significance of the experimental results and the adequacy of the resulting model to the real process.

ES models are graphically interpreted as triangles (v_1, v_2, v_3 – variable specific surface or fractional composition) on a cube of three prescription factors, applied to obtain new scientific information about the structure-forming role of mineral fillers and organic additives. The formation and development of space-time structures in open disperse systems is determined by the irreversibility of processes and associated with phase transitions of components when the system reaches some critical parameters [8, 9].

The task of analyzing the structure is quite complex, characterized by a set of various structural parameters. This complexity is due to the diversity and inconsistency of the development of elementary physical and chemical processes. The interactions of these processes in the future determine the level of properties of composite materials [10, 11]

Open disperse systems, which are characterized by a nonlinear synthesis of hydrate formation, are characterized by the effects of synergistic interactions [12, 13]. ES modeling allows us to assess the degree of interaction between different types of additives used in research.

4. Analysis of experimental results on experimental statistical models

At this stage, the change in water resistance and compressive strength of the composite gypsum binder is described and analyzed.

ES models of the influence on the properties of composite gypsum binder additives of various types and purposes, taking into account the interactions between them, are calculated in the PC COMPEX program [14, 15]. The program implements a sequential regression analysis with an experimental error $S\{Y\}$ of not more than 0.04 at risk degree $\alpha = 0.02$.

Compressive strength. The full ES model (2), which describes the effect of the studied factors on compressive strength, has the following form:

$$Y(f_{cm}) = \underbrace{\begin{array}{|c|}\hline +13.44v_1+3.43v_1v_2 \\ +11.96v_2\pm0v_1v_3 \\ +13.07v_3\pm0v_2v_3 \\ \hline\end{array}}_{(a)} + \underbrace{\begin{array}{|c|}\hline \pm0v_1x_4-0.71v_1x_5-0.93v_1x_6 \\ -0.79v_2x_4-0.61v_2x_5+0.70v_2x_6 \\ -1.45v_3x_4-0.62v_3x_5-0.87v_3x_6 \\ \hline\end{array}}_{(b)} + \underbrace{\begin{array}{|c|}\hline \pm0x_4^2\pm0x_4x_5 \\ -0.68x_5^2-0.62x_4x_6 \\ -1.09x_6^2\pm0x_5x_6 \\ \hline\end{array}}_{(c)} \quad (2)$$

The graphical interpretation of the model (2) is presented in Figure 1 in the form of mixed triangles in coordinates (v_1, v_2, v_3) on a quadratic diagram in coordinates (x_4, x_5). The content of additive-superplasticizer by Sika (x_6) was fixed at 0.1%; 0.6%; 1.1% levels.

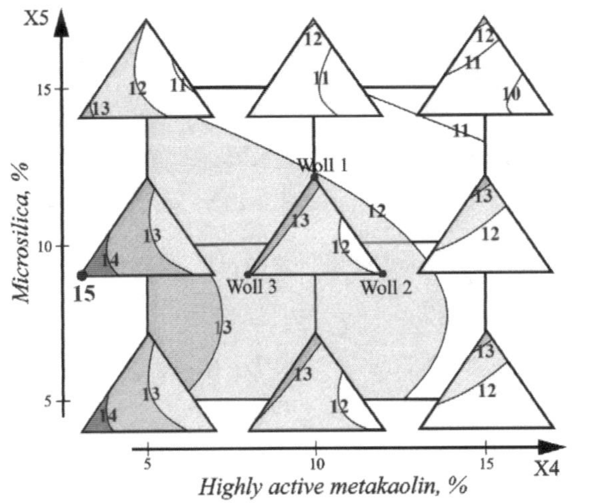

Figure 1. The effect of polymineral micropuzzolanic additives, micro-reinforcing filler additives (wollastonite) on the compressive strength of a composite gypsum binder.

The effect of the three fractions of wollastonite is illustrated on graphical images of ES models in the form of triangular diagrams. The needle-shaped form of wollastonite grain determines the main direction of its use as a micro-reinforcing filler. The physico - chemical affinity of wollastonite with composite building materials containing cement promotes active selective adsorption of binder hydration products, has a significant effect on rheological parameters, structure formation, strength and deformation properties of hardened composites.

Triangular diagrams show the effect on the properties of three dependent factors (v_1, v_2, v_3). These are dependent factors that are interconnected. The total content of all three components (Woll 1, Woll 2, Woll 3) remains constant, i.e. ($v1 + v2 + v3 = 1$). Triangular diagrams located on the square of the analyzed property, in this case f_{cm}, allow you to choose the best fractional composition of wollastonite, the content, the ratio of two or three fractions to each other.

The effect on the properties of micropuzzolanic filler additives is graphically interpreted as isolines on the square of two factors $x_4 \rightarrow$ (highly active metakaolin), $x_5 \rightarrow$ (microsilica). The content of highly active metakaolin as well as microsilica varied experimentally in the range of (10 ± 5) %.

Micropuzzolanic filler additives are highly pozzolanic. Microsilica has pozzolanic activity of 350 – 450 mg of bound lime per 1 g of microsilica. Highly active metakaolin, due to the content of active alumina, is able to bind a significantly larger amount of lime than silica fume. Its activity reaches more than 1000 mg of bound lime per 1 g of highly active metakaolin. The introduction of these additives into the composition of the composite gypsum binder should provide early and prolonged strength.

The high-quality superplasticizer on polycarboxylate base by Sika Visco Creit, which consists of specially synthesized chemically pure substances, was introduced to improve workability, reduce water demand and slow down the setting time of the composite gypsum binder.

As follows from the diagram in Figure 1, the maximum value of compressive strength is equal to f_{cm} = 14.5÷15.0 MPa. The obtained strength value is two to three times higher than the applied gypsum binder grade G-5. Portland cement PC-500-D0, non-additive in the amount of 21% from ∑(G+PC), was used as the second component of the composite gypsum binder.

The obtained maximum values of $f_{\rm cm}^{\rm max}$ are achieved when the content of Sika plasticizer in the amount of 0.1 – 0.6% (from b.m.), the minimum content of highly active metakaolin = 5% and the content of microsilica in the amount of 5 – 10% (from b.m.). Woll 3 wollastonite has a positive effect on f_{cm}. The appending of wollastonite in the mixture increases f_{cm} from 13 – 13.5 to 14.5 – 15.0 MPa, i.e. by 10 – 12.5%. Moreover, an increase in the content of Sika additive more than 1% reduces $f_{\rm cm}^{\rm max}$ by ~ 12% to 12 – 13 MPa.

Water resistance. At the next stage, an analysis of the change in water resistance by the value of the softening coefficient K_p under the influence of micropuzzolanic additives, micro-reinforcing modifier additives and Sika superplasticizer was carried out. The range of variation of all additives is limited by the experimental plan.

The change in water resistance of composite gypsum binder under the influence of the above additives is described by the model (3):

$$Y(K_p) = \underbrace{\begin{array}{|c|}\hline +0.676v_1+0.209v_1v_2 \\ +0.711v_2+0.432v_1v_3 \\ +0.690v_3-0.437v_2v_3 \\ \hline\end{array}}_{(a)} + \underbrace{\begin{array}{|c|}\hline \pm0v_1x_4-0.033v_1x_5+0.069v_1x_6 \\ +0.071v_2x_4-0.037v_2x_5\pm0v_2x_6 \\ +0.027v_3x_4-0.071v_3x_5-0.071v_3x_6 \\ \hline\end{array}}_{(b)} + \underbrace{\begin{array}{|c|}\hline \pm0x_4^2-0.057x_4x_5 \\ -0.033x_5^2\pm0x_4x_6 \\ +0.056x_6^2\pm0x_5x_6 \\ \hline\end{array}}_{(c)} \quad (3)$$

A graphical interpretation of the model in the form of two diagrams (triangles on a square) with a fixed content of additives Sika in the amount of 0.6 and 1.1% is presented in Figure 2.

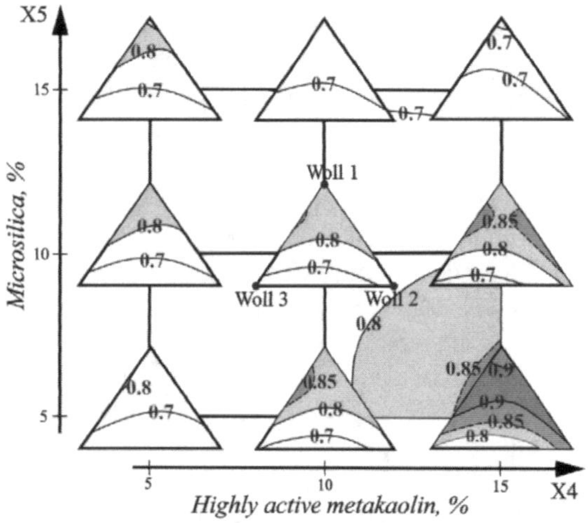

Figure 2. The effect of polymineral micropuzzolanic additives, micro-reinforcing filler additives (wollastonite) on the water resistance of a composite gypsum binder

The softening coefficient varies in the range from 0.6 to 0.92. Most of the compositions of 24 series of samples are waterproof, because their softening coefficient is 0.8 – 0.92. The required conditions for research problems softening coefficient value obtained at the maximum $K_p \geq 0.85$ content metakaolin high activity in an amount from 10 – 15% m. c. microsilica content 5 – 10% of m.v. and subject to the content in the mixture of wollastonite a certain fraction. The fractional composition of wollastonite for different mixtures is different for different contents of Sika superplasticizer. The compositions, which in the framework of the experiment have the maximum values of water resistance K_p = 0.9 – 0.92, were obtained when 15% of highly active metakaolin and 5% of microsilica were added to the composition of composite gypsum binder. The content of Sika is 0.1 or 1.1% by weight. The choice of the fractional composition of wollastonite remains the task of the technologist and is assigned taking into account the levels of other properties and the conditions of the multicriteria optimization problem [16].

As follows from the analysis of the diagrams (Fig. 1-2), compositions that have a maximum compressive strength of 14.5 – 15.0 MPa are characterized by low water resistance K_p = 0.7 – 0.75. Compounds that have a maximum water resistance of K_p = 0.85 – 0.92 have f_{cm} = 9.5 – 12.5 MPa. In accordance with the purpose of the study, for further analysis of the properties, samples with $K_p \geq 0.85$ were selected. The strength of samples with such values of the softening coefficient is equal to f_{cm} = 9.5; 10.5; 10 – 12.5 MPa with the content of Sika superplasticizer in the amount of 0.1%, 0.6% and 1.1%, respectively.

The values of the softening coefficient $K_p \geq 0.9$ have compositions that contain superplasticizer Sika = 1.1%, highly active metakaolin = 15%, microsilica = 5%. That is, the specific surface area of the microreinforcing additive of wollastonite (Woll), acting as a structure modifier, is an effective factor that allows you to adjust the properties of composite mixtures in a wide range and should be assigned taking into account the content of microfillers.

The choice of the wollastonite fraction remains the task of the technologist. It is possible to use monofraction wollastonite (Woll 2 = 1), bifraction wollastonite (Woll 1 + Woll 3 = 1) or (Woll 1 + Woll 2 = 1), or polyfraction wollastonite (Woll 1 + Woll 2 + Woll 3 = 1). The choice of the fraction of wollastonite may be due either to the technological capabilities of the preparation of mixtures, or to indicators of the quality of other properties.

5. Analysis of the effect of water repellents on the water resistance of a composite gypsum binder.

At the next stage of the study, the possibility of a further increase in water resistance due to volume and surface water repellents was analyzed.

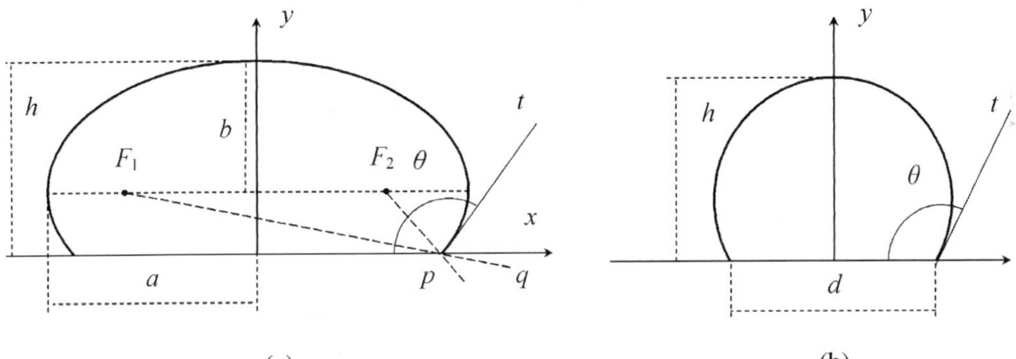

Figure 3. The shape of the droplets lying on a flat hydrophobic surface: (a) – "heavy" drop; (b) – "small" drop; θ – limiting contact angle, t – tangent to the meridian section of a drop at the point of separation of the three phases.

To evaluate the effect of water-repellent primers on the surface properties of samples from a gypsum binder, we measured the wetting angle by the method of a lying drop (sitting drop method). The shape of a drop resting on a horizontal flat surface in a gravitational field (Fig. 3(a)) is axisymmetric and is described by the Young – Laplace equation [17]:

$$\rho g y + \sigma \cdot (K(y) - K(h)) = 0, \quad (4)$$

where ρ is the density of the liquid, g is the acceleration of gravity, σ is the surface tension of the liquid (specific free energy of the liquid-gas interface), $K(y) = (R_1(y))^{-1} + (R_2(y))^{-1}$, $R_1(y)$ и $R_2(y)$ – principal radii of curvature of the surface of the drop at the ordinate point y, $K(h) = K(y)|_{y=h}$, i.e. on top of a drop.

Equation (4) has no analytical solution; therefore, numerical methods are used to calculate the droplet shape with high accuracy and to find the contact angle.

Small droplets under the action of capillary pressure are able to maintain a spherical shape (see Figure 3b). In this case, the contact angle can be easily found from elementary geometry θ. A drop can be considered small if its weight is much less than capillary pressure; the radius of a small spherical droplet r must satisfy the dependence $r << \left(\frac{2\sigma}{g\rho}\right)^{1/2}$ (here the gas density is considered negligible compared to the density of the liquid). Estimates show that this approximation is applicable to water droplets ($\sigma \approx 0.072$ N/m) with a radius of about 0.4 mm or less.

For larger droplets of radius $r \leq \left(\frac{2\sigma}{g\rho}\right)^{1/2}$ (up to 3.5 mm) their meridian section can be approximately considered as a segment of a conical section - an ellipse [18]. In this case, measuring the sizes of the semiaxes of the ellipse a and b (see Fig. 3(a)), it is easy to determine the positions of its characteristic point F_1 and F_2, then find the equations of the focal radii p and q connecting the foci of the ellipse with the point of separation of the three phases. The bisector of the external angle formed by the intersection of these focal radii will be the tangent t to the meridian section of the drop at the interface between the three phases. The angle of its inclination, which is easy to calculate, knowing the equations for p and q, is complementary to the desired angle θ.

In the experiment, the size of the liquid droplets is determined by the inner diameter of the dropper channel. In our case, it was 0.53 mm (with an outer diameter of the needle 0.8 mm). The masses of droplets of distilled water produced by the dropper were measured on an analytical balance. The needle used formed droplets with an average weight of 15.6 mg. The radius of the corresponding spherical drop is approximately equal to 1.55 mm, which, obviously, exceeds the criterion of smallness of the drop. To analyze the shape of such a drop, we decided to use the conical section method.

For photoregistration of the droplet shape on the test surface the CANYON CNR WCAM820 web camera with a resolution of 1600 × 1200 pixels was used. The standard camera lens was replaced with an optical system composed on the principle of a microscope. The field of view of the frame was about 9 mm, the image scale was determined using an object micrometer and was 5.75 μm / pixel. Two LED illuminators with brightness controls made it possible to combine oblique reflected and transmitted lighting to establish the optimal image contrast at the phase boundaries. During working with water-absorbing samples, the camera switched to video mode with a resolution of 800 × 600 pixels and with a frame interval of 100 ms.

The image obtained by the camera was saved on a computer and analyzed using a graphical editor. In the graphical editor an ellipse that satisfactorily coincided with the contour of the drop was selected. After that, the sizes of its semiaxes, the coordinates of the center and the coordinates of the points of separation of the three phases was read, then the contact angle was calculated by the method described above.

Figure 4 shows a drop of distilled water on the surface of a sample that has not been primed. The surface of the sample is hydrophilic, as evidenced by a contact angle of less than 90°. Water spreads and is absorbed by a gypsum binder: the drop on the figure shows at various time intervals that have passed since the moment of contact.

Water-absorbing surfaces can be characterized, in addition to the contact angle, by the rate of absorption through a single contact surface. The absorption rate can be estimated by the decrease in droplet volume over time, if the evaporation rate can be neglected, which is acceptable at a high absorption rate.

The calculation of the droplet volume was carried out under the assumption that the droplet is a truncated ellipsoid of rotation. It is easy to show that in this case the volume V is determined by the height of the droplet h and the dimensions of the semiaxes of its meridian section:

$$V = \pi a^2 \cdot \left(h - \frac{b}{3} + \frac{(b-h)^3}{3b^2}\right) \quad (5).$$

(a) (b) (c)

Figure 4. A drop of water on the surface of the untreated sample after the onset of contact: (a) – 0.4 s ($\theta = 83°$); (b) – 2.0 s ($\theta = 72°$); (c) – 5.0 s ($\theta = 62°$)

The experiments showed that the droplet absorption rate into the samples can be considered high, because during the time of complete absorption, the volume of the control drop on the metal surface remained unchanged within the measurement error.

On untreated samples, the contact angle decreases during the first two seconds of contact from 86° to 72°, but the drop volume remains unchanged within the measurement error, therefore, only the drop spreads. The start of absorption is recorded by us from the 3rd second. The maximum absorption rate is 5–6th second and amounts to 0.017 $g \cdot cm^{-2} \cdot s^{-1}$. The total absorption time of the droplet varied in the range of 12-30 s with an average contact area of 0.15 cm^2.

Figure 5 are shown drop of distilled water on the sample surface treated with a hydrophobic primer. The surface of the sample acquired hydrophobic properties: the contact angle for water exceeds 90°. On these surfaces was not observed the spreading and absorption of a drop.

Figure 5. A drop of water on a sample surface treated with a hydrophobic prime

The average contact angle for gypsum surfaces treated with PGK-12S and PGK-24 primers is, respectively 129 ± 7° and 130 ± 5°. Taking into account the specified error, these contact angles of wetting can be considered the same, thus, primers PGK-12S and PGK-24 demonstrate the same efficiency in relation to hydrophobization of the gypsum binder surface. A slightly worse result is given by the WHITE primer, for which the average contact angle was 121 ± 4°.

The hydrophobizing primer introduction into the volume of a gypsum binder at the mixing stage without subsequent sample surface treatment leads to a different result. Fig. 6 shows the evolution of a distilled water drop on the surface of such a sample.

Immediately after the drop hits the surface and stabilizes its shape, the contact angles of wetting take on values exceeding 90°, which is typical for hydrophobic surfaces. The drop does not spread on the surface, and the area of the contact spot does not increase. However, the drop is absorbed into the sample volume. As the droplet volume decreases while maintaining the area of the contact spot, the contact angle decreases, and after a short time takes on the values characteristic of hydrophilic surfaces.

(a) (b) (c)

Figure 6. A drop of water on the surface of the sample with WHITE primer added to the volume of the binder after the onset of contact: (a) – 1.0 s ($\theta = 122°$); (b) – 3.9 s ($\theta = 90°$); (c) – 7.0 s ($\theta = 81°$)

6. Conclusions

An analysis of the results of the study allows us to conclude that the proposed composition of the multifunctional modifier allows to obtain a waterproof modified composite gypsum binder with a softening coefficient $K_p \geq 0.9$ on gypsum grade G-5, rather than gypsum grade G-10 used for concrete.

It was shown that the complex modification of a composite gypsum binder on gypsum of the G-5 grade and Portland cement PC500-D0 with micro-micropozzolanic and micro-reinforcing filler additives and Sika superplasticizer in optimal proportions allow to obtain a composite gypsum binder with a softening coefficient of 0.9. That is, the specific surface area of the micro-reinforcing wollastonite additive, acting as a structure modifier, is an effective factor that allows you to adjust the properties of composite mixtures over a wide range and should be assigned taking into account the content of micro-fillers.

Volumetric hydrophobization provides an additional increase in compressive strength and water resistance. Further studies are aimed at obtaining and optimizing multifunctional modifiers with a wide spectrum of action.

References

[1] Dvorkin L I 2019 Modified water-cement ratio rule for the design of air-entrained concrete. *Magazine of Civil Engineering* **85(1)** 123–135

[2] Khaliullin M I and Gaifullin A R 2010 Plaster dry mixes on the basis of composite gypsum knitting the raised water resistance. *News of the KSUAE.* **2** 292-296

[3] Shinkevich E S, Linnik D S and Ysypchuk V I 2019 Improvement of Arbolit Concrete Bio-Resistance with Additives by Nanomodificators *Materials Science Forum* **968** 76-8

[4] Tymnyak A B, Linnik D S, Shinkevich E S and Ysypchuk V I 2015 Optimization of compositions and properties of fine-grained concrete on a composite binder. *Bulletin INCERCOM sci. res. inst. of constr.* **6** 136-140

[5] Derevianko V N, Kondratieva N V and Hryshko H M 2018 Study of impact factors and the mechanism of process of gypsum binder hydration in the presence of nanomodifiers. *Journal of Chemistry* **6** 92-100

[6] Derevianko V N, Kondratieva N V, Sanitskiy N A and Hryshko H M 2018 Impact of granulometric composition on mineral binder hydration processes. *Journal of Engineering Science* **3** 74-84

[7] Derevianko V N, Kondratieva N V, Sanitskiy N A and Hryshko H M 2016 Investigation of hydration processes of nanomodified calcium sulfate hemihydrate. *Jour. of Tech. University of Moldova and Moldavian Engineering Association* **4** 37-41

[8] Derevianko V N, Kondratieva N V, Hryshko H M and Sanitskiy M A 2020 Modelling the Mechanism of Mineral Binders Hydration Processes in a Macro-Micro-Nanosystem. *Nanosystems, nanomaterials, nanotechnologies.* **18** 107-124

[9] Vinnichenko V, Gabitov A, Salov A, Gaisin A and Kuznetsov D 2018 The heat loss calculating methods of external walls in the buildings reconstruction *MATEC Web Conf.* **Vol 230** (2018)
[10] Derevianko V N, Maksimenko A A and Hryshko H M 2015 Study of polymer-modified powder dry mix produced from magnesia-bishofit composition with additives. *Jour. of Tech. University of Moldova and Moldavian Engineering Association* **4** 35-39
[11] Vinnichenko V, Ryazanov A and Krot O 2018 Construction binders and environmental indicators of their production *MATEC Web Conf.* **Vol 230** (2018)
[12] Kondratieva N Barre M Goutenoire F Sanytsky M 2017 Study of modified gypsum binder. *Construction and Building Materials* **149** 535–542
[13] Kondratieva N, Barre M, Goutenoire F, Sanytsky M and Rousseau A 2020 Effect of additives SiC on the hydration and the crystallization processes of gypsum. *Construction and Building Materials* **235** pp 117479
[14] Shinkevich E S, Tymnyak A B, Linnik D S and Tertychny A A 2013 Composition optimization of dry building mixtures based on experimental-statistical models. *Build. materials prod. and sanitary ware* **48** 179-183
[15] Shinkevich E S, Tymnyak A B, Linnik D S and Ysypchuk V I 2014 *Composites optimization modeling* (Odessa: Astroprint) 106-108
[16] Shinkevich E S, Linnik D S and Ysypchuk V I 2015 The effect of highly active pozzolanic additive on the properties of composite gypsum binder and arbolite concrete based on it. *Bulletin of Odessa State Academy of Civil Engineering and Architecture* **57** 273-278
[17] Finn R 1986 Equilibrium capillary surface. (NY: Springer-Verlag) 245 p
[18] Whyman G and Bormashenko E 2009 Oblate spheroid model for calculation of the shape and contact angles of heavy droplets. *Journal of Colloid and Interface Science* **331(1)** 74-177.

Fiber type influence on the reinforced concrete under axial tension

M Surianinov[1,2], D Kirichenko[1], I Korneieva[1], S Neutov[1]

[1] Odessa State Academy of Civil Engineering and Architecture, 4, Didrikhsona st., Odessa, 65029, Ukraine

[2] Address all correspondence to this author. E-mail: sng@ ogasa.org.ua

Abstract. The results of experimental studies of properties of steel fiber-reinforced concrete of different types under axial tension are presented. A stand with a flexible design of the grip element connection has been developed, which provides axial load application and minimizes the effect of stress concentration at the ends of the specimen. The design is based on the ideas of the well-known Michaelis device. The design of the stand is such that the distance between the axis of load application and the central joint is 108 cm and between this joint and the axis of the test specimen is 21 cm, resulting in the load applied to the specimen being 5.143 times the applied load. The limits of fiber concrete axial tensile strength have been determined - 1.28 MPa when reinforced with wave or spun fiber and 1.37 MPa when reinforced with anchor fiber or a combination of the three types of fiber, which is 4.1 % and 4.4 % of compressive strength, respectively. It was also found that the concrete reinforced with anchor fiber or with the combined dispersed fiber reinforcement has higher deformation properties than the concrete reinforced with wave or spun fiber. The analysis of strength and deformation properties of fiber concrete when reinforced with different types of fibers allows to recommend anchored fibers for practical application.

Introduction

Concrete is known to perform poorly in tension. Its tensile strength is only 7-10% of its compressive strength, which is a significant disadvantage of concrete. There are several types of tensile strength of concrete: flexural tensile strength, axial tensile strength, splitting tensile strength.

In practical terms, the tensile strength of concrete shows the ability to withstand various deformation loads while maintaining the geometry of the structure unchanged. In addition, the tensile strength of concrete shows the resistance of the material to temperature variations and is also taken into account when calculating the load-bearing capacity of structures, especially those working in flexure.

The axial tensile strength is the most difficult to determine. One of the options is the tension of dog-bone specimens. The bending tensile strength is determined on 100x100x400 mm prisms; 150x150x 600 mm, etc. Samples are tested under the action of two forces applied in 1/3 of the span. The destruction of concrete occurs from the tensile stresses, reaching the greatest values in the lower layer of the tensile zone. Due to the complexity of determining the pure tensile and flexural strength of concrete, the determination of the tensile strength of concrete in splitting (sometimes this method proposed by F. Carneiro is called the Brazilian test) is widely used. The standard cubes used are split on presses with steel or plywood liners. Splitting strength is slightly higher than pure tensile strength, on average 30%. For conventional concretes, even in bendable structures, the tensile strength of concrete is not regulated. Tensile stresses in them are taken up by the reinforcement. In the concrete of the tensile zone due to its low extensibility cracks are allowed and formed. But the width of their opening is limited (in the limit -

up to 0.3 mm). At the same time, cracks are inadmissible in a number of structures - road surfaces, tanks, hydraulic structures. For them, the tensile strength becomes essential and may be the main standardized characteristic.

One way to increase the tensile strength and cracking resistance of concrete is its dispersed steel fiber reinforcement.

1. Materials and methods

Determination of physical, physical-mechanical and construction-technical properties of concrete and mixtures with dispersion reinforcement has been carried out in accordance with the current regulatory documents and generally accepted methods. Mathematical methods of experiment planning and analysis of the obtained experimental results by methods of mathematical statistics have been used.

2. Analysis of publications

The tensile strength of concrete and fiber concrete has been the subject of study by many researchers. Wu et al. [1] conducted experimental studies on the tensile strength of concrete at low strain rates using three different test methods: axial stretching, splitting stretching and flexural stretching. In the latter case, the ultimate tensile strength was the greatest. The effect of specimen end grips on the ultimate tensile strength of concrete was investigated by Wille et al. [2], Li Z. et al. [3] and Zijl et al. [4].

A number of works [5-7] investigated the axial tensile strength of lightweight concrete samples with different aggregate sizes. The authors found that the tensile strength increased with decreasing specific weight of the concrete.

It has been noted that the end conditions affect crack formation and strain distribution in the cross section of the tested specimens [8].

I.K. Mohammad investigated the axial tensile strength and compared the results with the bending tensile strength. They were comparable [9]. V. Sarfarazi and colleagues conducted a Brazilian test to compare the results of the same two methods. The test results showed that the axial tensile strength was much lower than in the Brazilian test - the difference was about 33% [10].

However, all of these methods have some kind of disadvantages - inhomogeneous fracture plane, stress concentration and slippage at the end of specimens, and bending loads due to imprecise specimen alignment during testing.

3. Problem formulation

Thus, the tensile strength of concrete and fiber concrete is significantly affected by a number of factors - the test method, the design of the test bench, the nature of the load application, the composition of the mixture, etc., so the problem of determining the tensile strength of fiber concrete continues to be relevant.

4. Aim of Paper

The purpose of this work is to experimentally investigate the properties of steel fiber concrete reinforced with steel fibers of different types under axial tension, in order to further ensure the specified performance characteristics of steel fiber concrete structures.

5. Research results

In order to make samples in the process of kneading the concrete mixture, fibers were evenly added, the total volume of which was 1% of the product itself. This volume was determined as optimal in previous studies [11]. The tests were conducted on prisms and cubes with dimensions of 100x100x400 mm and 100x100x100 mm, respectively, cured for 28 days. The prisms and cubes were made in accordance with regulatory documents [12]. Four groups of prisms and cubes were made. One of each - from ordinary concrete (with coarse aggregate dimensions up to 10 mm) class C20 / 25, and three more - with different types of fibers (anchored, spun and corrugated, figure 1). Each group consisted of six samples.

The fiber used for the research was provided by Stalkanat-Silur and made in accordance with the requirements of EN 14889-1:2006 [13].

In all experiments, cement grade 400 and washed river sand were used. The water-cement ratio was 0.449.

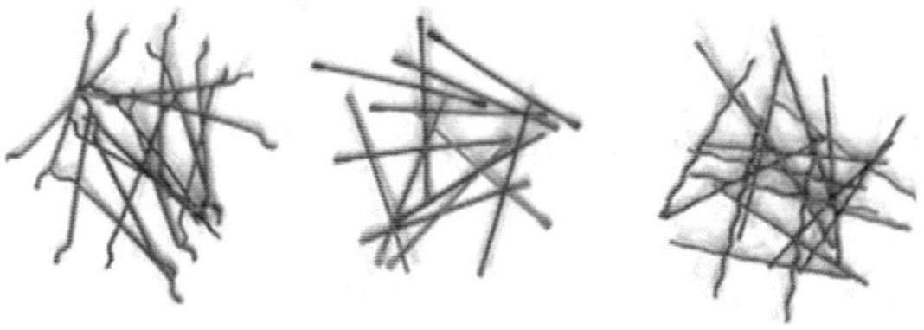

Figure 1. Steel fibers of three types.

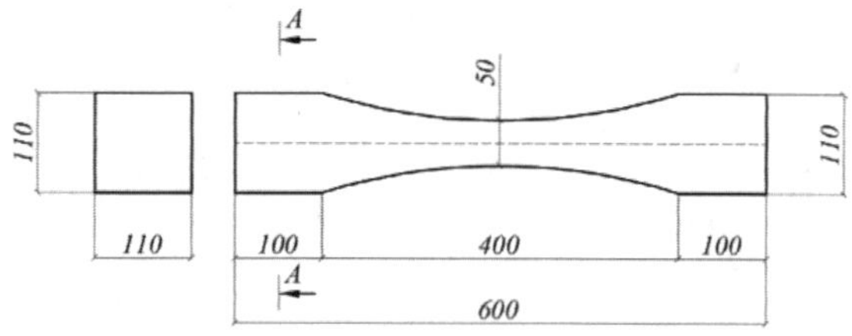

Figure 2. Tensile test specimen

At the second stage, 32 dog-bone specimens with 1.0% dispersed reinforcement were made (8 specimens each of concrete reinforced with anchor, spun and wave fibers, and another 8 of concrete reinforced with a mixture of three types of fibers in equal amounts). All specimens had the same dimensions, a cross-section of 110 x 110 mm at the ends and a cross-section of 50 x 70 mm in the middle (figure 2).

For the tests a specially designed test stand with a flexible design of the connection of the gripping elements, which provides axial application of the load and minimizes the effect of stress concentration at the ends of the specimen (figure 3). The design is based on the ideas of the well-known Michaelis instrument, which is usually used to determine the bending strength of binders in the form of standard beam specimens.

The design of the bench is such that the distance between the axis of load application and the central joint is 108 cm, and between this joint and the axis of the test specimen is 21 cm. It follows that the load applied to the specimen is 5.143 times the applied load.

Figure 3. Test specimen.

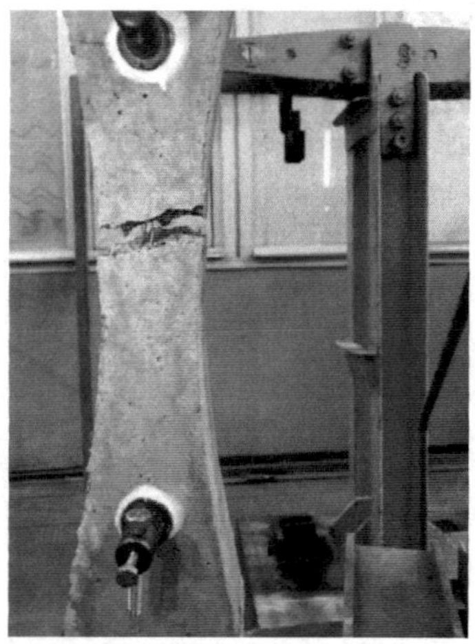

Figure 4. Destruction of the sample.

The division value of the dynamometer when fixing the load transferred to the sample is 1 division = 0.3 kg. The base of the indicators attached to the dog-bone specimen (figure 3) was 24 mm. The load was applied in steps, with an endurance at each step of loading.

After processing the test results of each series by the methods of mathematical statistics [14, 15], the average values of the controlled values were obtained.

Table 1 shows the average values of relative strains at each test step.

Table 1. Relative deformations at the four reinforcement options

Fiber type							
Anchor		Crimped		Wave		Mix	
Load, Kg	Strain, $\varepsilon\cdot10^4$	Load, Kg	Strain, $\varepsilon\cdot10^4$	Load, Kg	Strain, $\varepsilon\cdot10^4$	Load, Kg	Strain, $\varepsilon\cdot10^4$
0.00	0.0000	0.00	0.000	0.00	0.000	0.00	0.000
108.00	0.083	123.43	0.042	123.43	0.042	108.00	0.167
154.29	0.167	169.71	0.083	169.71	0.083	154.29	0.208
200.57	0.208	216.00	0.167	216.00	0.125	200.57	0.250
200.57	0.208	216.00	0.167	216.00	0.125	200.57	0.250
246.86	0.292	262.29	0.208	262.29	0.167	246.86	0.333
246.86	0.333	262.29	0.208	262.29	0.167	246.86	0.333
293.14	0.375	308.57	0.292	308.57	0.250	293.14	0.438
293.14	0.375	308.57	0.292	308.57	0.250	293.14	0.438
339.43	0.500	354.86	0.333	354.86	0.333	339.43	0.542
339.43	0.500	354.86	0.333	354.86	0.333	339.43	0.542
385.71	0.625	401.14	0.375	401.14	0.458	385.71	0.646
385.71	0.625	401.14	0.375	401.14	0.458	385.71	0.646
432.00	0.833	447.43	0.417	447.43	0.542	432.00	0.708
432.00	0.833	447.43	0.458	447.43	0.583	432.00	0.708
478.29	1.042	-	-	-	-	478.29	0.750
478.29	1.083	-	-	-	-	478.29	0.777

The load-strain diagrams for the four reinforcement options are shown in figure 5, 6.

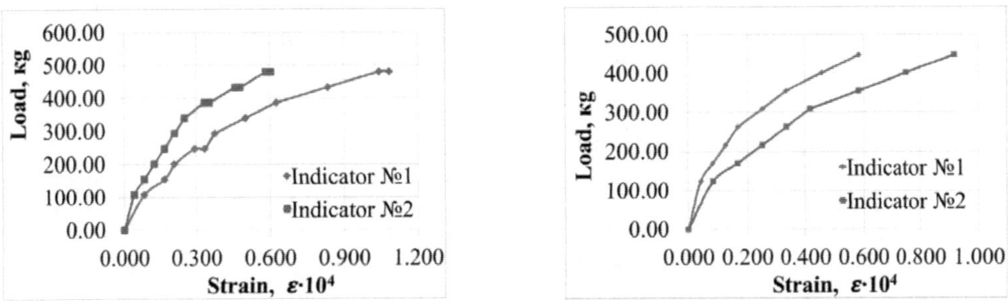

Figure 5. Load-strain diagrams Anchor, Crimped fibers.

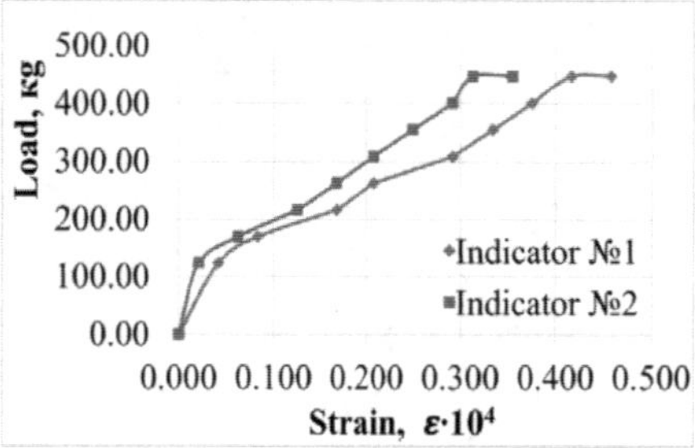

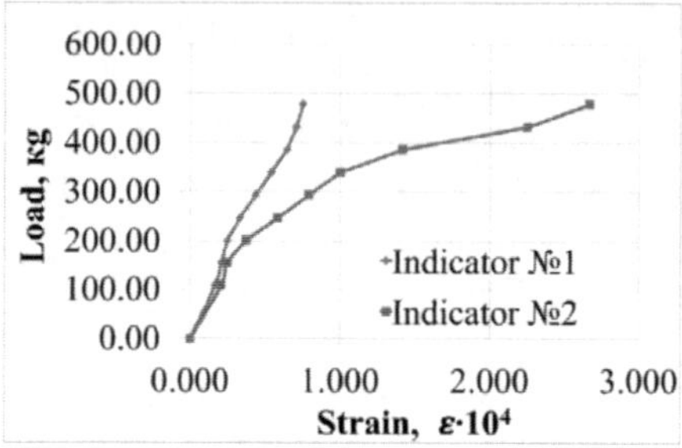

Figure 6. Load-strain diagrams Wave, Mix fibers.

At the rupture, the average number of fibers caught in the section of the dog-bone specimen was 16 units, the anchor fiber - 22 units, the spun fiber - 10 units. And at the rupture of the dog-bone specimen with combined reinforcement the number of fibers was: anchor fiber - 0 units, spun fiber - 19 units, wave fiber - 2 units.

6. Conclusions

Thus, the design of the test stand for tensile testing of concrete and fiber concrete dog-bone specimens, which provides an axial load application and minimizes the effect of stress concentration at the ends of the specimen, was developed. The design is based on the ideas of the Michaelis instrument, which is usually used to determine the flexural strength of binders in the form of standard beam specimens.

At the first stage of tests it was found that the optimal characteristics of the fiber concrete mixture is a matrix with coarse aggregate ≤ 10 mm with 1.0% fiber reinforcement.

At the second stage, the limits of fiber concrete axial tensile strength were determined - 1.28 MPa when reinforced with wave or spun fiber and 1.37 MPa when reinforced with anchor fiber or a combination of the three types of fiber, which was 4.1% and 4.4% of the compressive strength, respectively.

It is also established that concrete reinforced with anchor fiber or with combined dispersed reinforcement has higher deformation properties than concrete reinforced with wave or spun fiber. The analysis of strength and deformation properties of fiber concrete when reinforced with different types of fibers allows to recommend anchored fibers for practical application.

References

[1] Wu S, Chen X, Zhou J 2012 Tensile strength of concrete under static and intermediate strain rates: Correlated results from different testing methods *Nuclear Engineering and Design* **250** pp 173-183

[2] Wille K, El-Tawil S, Naaman A E 2014 Properties of strain hardening ultra high performance fiber reinforced concrete (UHP-FRC) under direct tensile loading *Cement and Concrete Composites* **48** pp 53-66

[3] Li Z, Kulkarni S M, Shah S P New test method for obtaining softening response of unnotched concrete specimen under uniaxial tension *Experimental Mechanics* **33(3)** pp 181–188

[4] Zijl G, Slowik V, Filho R, Wittmann F, Mihashi H 2016 Comparative testing of crack formation in strain-hardening cement-based composites (SHCC) *Materials and Structures*. **49(4)** pp 1175 - 1189

[5] Mechtcherine V 2007 Testing behaviour of strain hardening cement-based composites in tension – summary of recent research *RILEM-Symposium on high-performance fibre reinforced cementitious composites RILEM PRO* **53** pp 13–22

[6] Elvery R H, Haroun W A 1968 A direct tensile test for concrete under long- or short term loading *Magazine of Concrete Research* **20(63)** pp 111-116

[7] Xie N, Liu W 1989 Determining tensile properties of mass concrete by direct tensile test *ACI Materials Journal* **86(3)** pp 214-219

[8] Ueda M, Hasebe N, Sato M, Okuda H 1994 Fracture mechanism of plain concrete under uniaxial tension *Concrete Library of the JSCE* **24** pp 31-45
[9] Mohammad I K 2011 Direct tensile strength measurement of concrete *Appl. Mech. Mater* **117–119** pp 9–14
[10] Sarfarazi V, Ghazvinian A, Schubert W, Nejati H R, Raouf H 2016 A new approach for measurement of tensile strength of concrete *Period. Polytech. Civ. Eng* **60 (2)** pp 199–203
[11] Neutov S F, Boyadzhi A A, Korneeva I B 2018 Determination of the main physical and mechanical characteristics of the steel fiber concrete mixture of the optimal composition *World science* **5 33 (2)** pp 26-30
[12] State standard of Ukraine B.V.2.7-214: 2009 [valid from 2009-12-22] 2010 *Concrete. Methods for determining strength according to control samples* Ministry of Regional Development of Ukraine
[13] EN 14889-1:2006 Fibres for concrete - Part 1: Steel fibres - Definitions, specifications and conformity
[14] Rykov V V, Ytkyn V Yu 2008 *Mathematical statistics and experiment planning* Russian State University of Oil and Gas named after I M Gubkin p 210
[15] Kuznetsova E V 2011 *Mathematical planning of the experiment* Perm State Technical University p 33

Experimental studies of road slabs

M Surianinov[1,2], S Neutov[1], I Korneieva[1], D Kirichenko[1]

[1] Odessa State Academy of Civil Engineering and Architecture, Didrikhsona st., 4, 65029, Odessa, Ukraine

[2] Address all correspondence to this author

E-mail: sng@ogasa.org.ua

Abstract. The results of experimental studies of bearing capacity and cracking resistance of road slab models made of steel-fiber concrete according to the span loading scheme are presented. The bearing capacity of the models has been fixed at 27.96 kN, which is 2 times higher than the load corresponding to the beginning of cracking (14.8 kN). At the moment of bearing capacity loss, the height of compressed zone was 1 cm, which is one third of its original size. The relative strains before the first crack appeared were 0.1175×10-4 and 0.235×10-4 in the compressed and stretched zones, respectively, and similar limiting relative strains reached values of 3.715×10-4 and 9.235×10-4. The maximum crack opening width was substantially less than in the same tests of the slabs without dispersion reinforcement and was 0.3 mm, which is explained by the more ductile properties of steel fiber concrete, which prevent crack opening. The final deflection in the middle of the plate model span was 10.4 mm, which is 5.5 times greater than the deflection before cracking began (1.9 mm).

1. Introduction

Road slabs are a good alternative to conventional asphalt pavement. They are used for the construction of temporary and permanent roads, airfields, warehouses and industrial sites, as well as where you want to get a durable surface that can withstand the weight of heavy machinery in the shortest possible time. Road slabs are also used in private construction, for example, for arrangement of yard area. A wide range of slabs allows you to choose the product with the most suitable performance characteristics.

The first road slabs appeared in the early twentieth century. It was an experiment in which it was wanted to compare the strength of conventional asphalt and concrete slabs. The experiment lasted for several decades, as it was necessary to compare the durability of the pavement. It turned out that asphalt needed to be replaced after 10 years, and every 3-4 years it had to be repaired. The concrete slabs lasted about 40 years without serious deformation [1]. All concrete slabs are characterized by high frost resistance, strength, resistance to rainfall and mechanical damage, high durability. The maximum load depends on what kind of concrete was used in the production. Slabs are produced in different sizes and shapes, different thicknesses, with different types of reinforcement and using different grades of concrete.

The undeniable advantage of concrete road slabs is the possibility of their multiple use. If the temporary roadway is no longer in use, its fragments are dismantled and transported to another place for subsequent laying of a new route. Used products are only visually different from the new fragments. The characteristics of durability remain unchanged, and the cost of recycled raw materials is 30%-40% lower than the price of new products. This makes it possible to reduce the cost of new roads without compromising their quality and durability.

2. Recent researches analysis

The general concept of calculation of road slabs was proposed by B. N. Zhemochkin and A. P. Sinitsyn [2]. Currently, regulatory documents on the calculation and design of road slabs developed in many countries. So in Ukraine it is DSTU B B.2.6-120:2010 [3], in European countries - EN 1992-1-1 (2004): Eurocode 2: [4]. Nevertheless, the interest of scientists to the problem does not weaken. So, in [5] the results of research of stresses in road slabs by polarization-optical method are stated. Experimental and theoretical studies of Belarusian scientists are given in [6, 7]. The evolution of precast concrete pavements in different countries and, in particular, in the USA is described in [8]. Here we present an overview of the results of accelerated loading tests with the help of heavy vehicle simulator. Extensive experimental studies to investigate the characteristics of steel fiber concrete road slabs (SFRC) are presented in [9]. The experimental results show that steel fibers significantly increase the bearing capacity and ductility of the slabs. Let us also note the works [10 - 12], which reflect various aspects of strength and deformability of road slabs.

3. The purpose of this work is to experimentally investigate the bearing capacity and fracture resistance of road slab models made of steel fiber concrete.

4. Materials and methods

The slab model design has dimensions in plan of 1375x825 mm and height of 60 mm. The series was made - three models of reinforced concrete, reinforced with Bp1 meshes, with the addition of steel fiber in the concrete mixture of 1% of the total volume of the product.

The laboratory tests were conducted in accordance with the current standards [13-15] according to the span loading scheme shown in Figure 1.

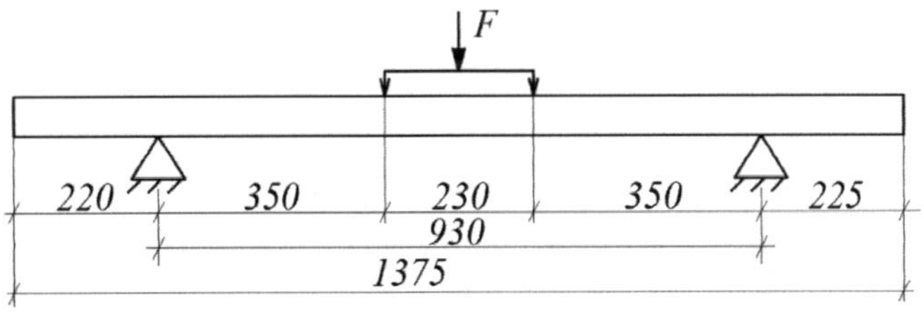

Figure 1. Testing scheme

On the upper and lower surfaces of the plate were mounted dial indicators with a base of 36 cm. The first four (1-4) - in the compressed zone, and two more (5, 6) - in the tensile zone (Figure 2). Load cells were also mounted - on the axis of symmetry, under indicators 2 and 4.

The load was applied in stages, and at each stage the longitudinal deformations were recorded according to the readings of indicators and strain gauges, as well as deflections by the Maximov deflectometers installed on both sides in the middle of the plate span.

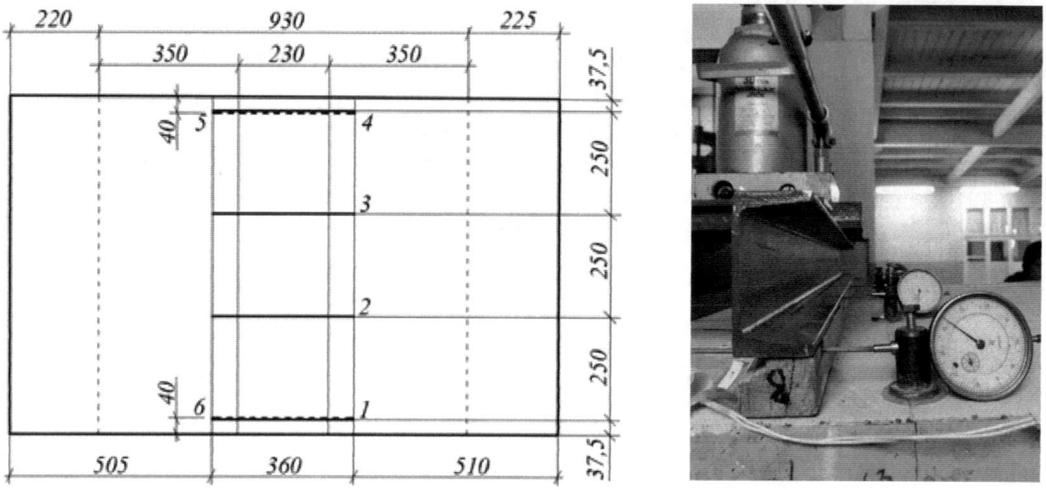

Figure 2. Indicator layout
a – scheme, b – indicator mount

5. Research results

The readings of all the instruments used on the three different models differed within the statistical error. Based on the data obtained, plots of relative strain vs. load were plotted, which can be used to trace the entire deformation process of the slab models (Figures 3-5).

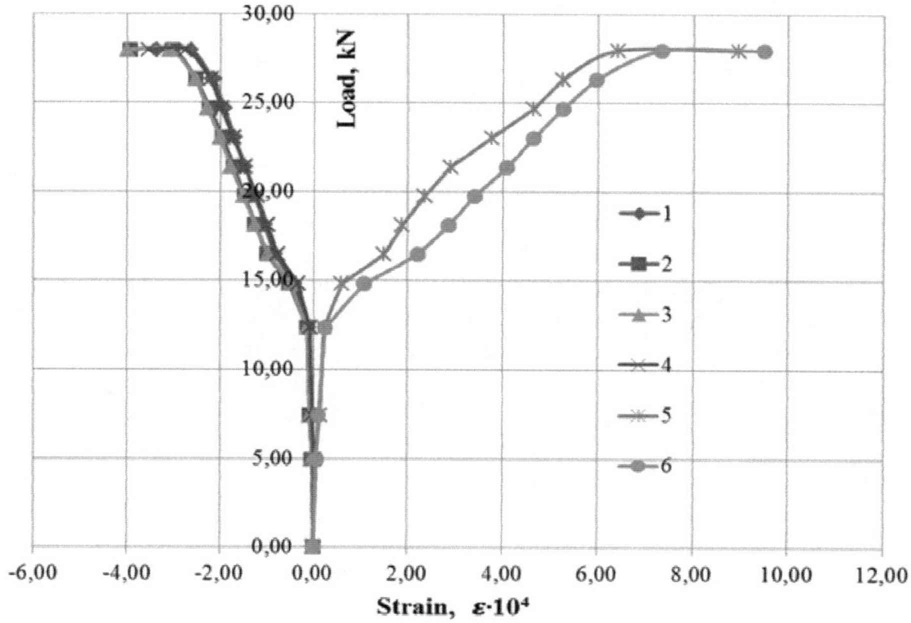

Figure 3. Relative deformation according to indicators

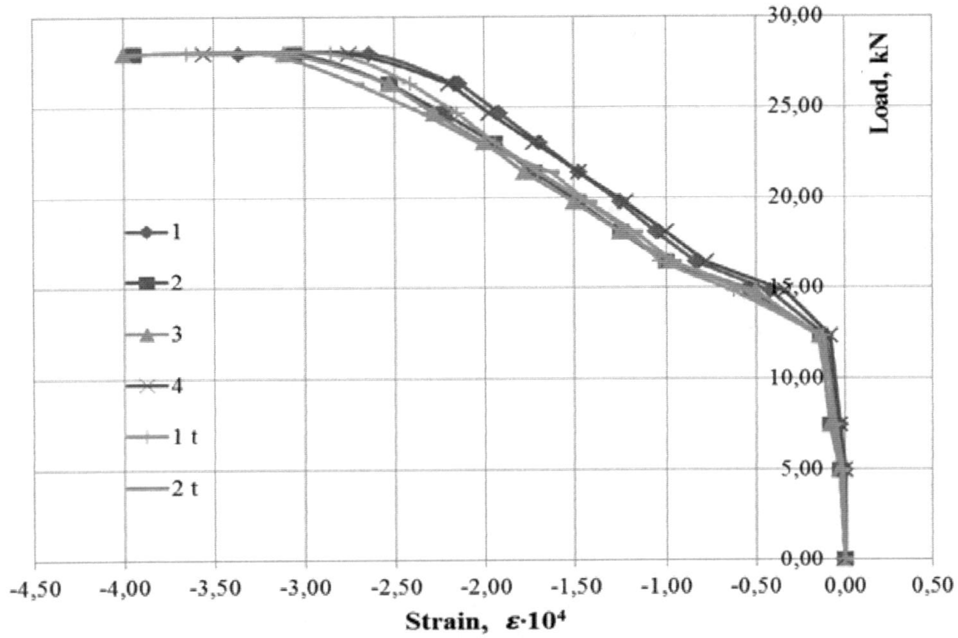

Figure 4. Relative strain in the compressed zone according to the readings of indicators and strain gauges

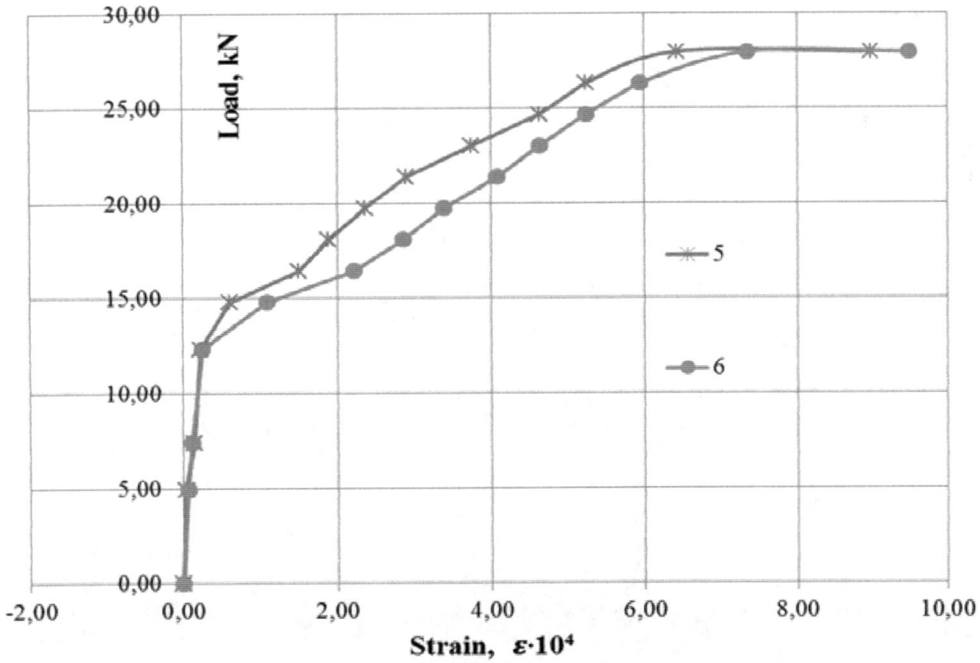

Figure 5. Relative strain according to the readings of indicators 5 and 6 in the tensile zone

The nature of the deflections change in the slab graphically and visually is shown in Figures 6, 7.

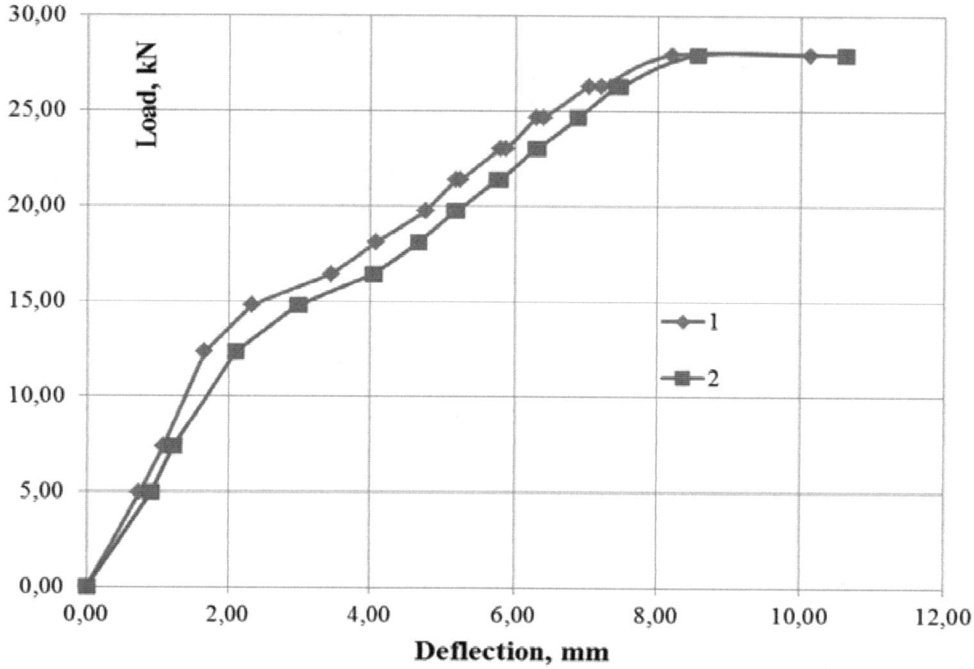

Figure 6. Slab deflections

Figure 7. Deflections in the span

The readings of the deflectors mounted in the middle of the plate span on both sides differed little from each other at each stage of loading, which testifies to the correctness of the load application, i.e. to the observance of symmetry. Before the appearance of the first crack, the values of deflections changed insignificantly, and starting from the load of 14.80 kN, deflections began to increase gradually, and starting from the load of 21.38 kN, these increments became significant, which is reflected in the graphs.

During the tests, the appearance of cracks and the nature of their opening were recorded. The width of the opening was measured using a Brinell microscope with a division value of 0.1 mm. The first crack formed in the middle of the plate's span at a load of 14.80 kN. At the moment of its appearance, its width of opening was 0.05 mm. At the load of 18,09 kN the second crack with the opening width of 0,1 mm was fixed. During exposure at the same stage of loading, the third crack with the opening width of 0.05 mm was formed. First, measurements of the crack opening width were carried out on the side faces, and then on the bottom surface of the slab models. Figure 8 shows the bottom view of one of the plates at the end of the tests. All the cracks were formed in the area of maximum bending moment. At the maximum load of 27.96 kN a width of the central crack opening of 0.3 mm was fixed.

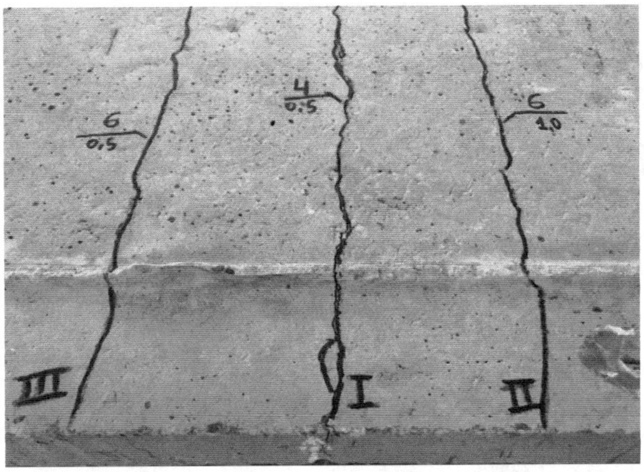

Figure 8. Cracks on the bottom surface of the slab

6. Conclusions

The conducted studies have fixed the bearing capacity of the road slab models tested by the flight loading scheme at 27.96 kN, which is 2 times higher than the load corresponding to the beginning of cracking (14.8 kN).

At the moment of loss of bearing capacity, the height of the compressed zone was 1 cm, which is one-third of its original size.

The relative strains before the first crack appeared were 0.1175×10-4 and 0.235×10-4 in the compressed and stretched zones, respectively, and similar ultimate relative strains reached values of 3.715×10-4 and 9.235×10-4.

The maximum crack opening width was significantly less than in the same tests of the slabs without disperse reinforcement and was 0.3 mm, which is explained by the more ductile properties of steel fiber concrete, preventing crack opening.

The final deflection in the middle of the slab model span was 10.4 mm, which is 5.5 times greater than the deflection before cracking began (1.9 mm).

References

[1] Radovskyi B S 2015 Cement concrete pavements in the USA - designing *Car roads* **2** pp 48-60
[2] Zhemochkyn B N, Synytsyn A P 1962 *Practical methods for calculating foundation beams and slabs on a firm base* Moscow p 239

[3] State standards of Ukraine V.2.6–120:2010 (HOST 21924.0–84, MOD) 2010 *Structures of buildings and structures. Reinforced concrete slabs for covering urban roads. Technical conditions* p 56
[4] EN 1992-1-1 2004 Eurocode 2 *Design of concrete structures* - Part 1-1: *General rules and rules for buildings* [Authority: The European Union Per Regulation 305/2011, Directive 98/34/EC, Directive 2004/18/EC]
[5] Shyrnyn Yu A, Cherniakevych V Y 2009 Investigation of the stresses of road slabs by the polarization-optical method *YVUZ Forest Journal* **6** pp 60-67
[6] Semeniuk S D, Kumashov R V, Ketner E A 2016 Bearing and operational capacity of reinforced concrete slabs of road cover *Science and construction* **3** pp 11-18
[7] Semeniuk S D, Kumashov R V 2018 Reinforced concrete slabs of road surface covering on elastic half-space *International Journal for Computational Civil and Structural Engineering* **14(2)** pp 149–157
[8] Houben L J M, Poot S, Huurman M, Kooij J Van Der 2004 Developments on the ModieSlab Innovative Concrete Pavement Concept *9th International Symposium on Concrete Roads* pp 4-7
[9] Sorelli L, Meda A, Plizzari G A 2006 Steel Fiber Concrete Slabs on Ground: A Structural Matter *ACI Structural Journal* Technical Paper **103 S58** pp 551-558
[10] Zhang Da Shan, Yu Li Dong 2011 Experimental Behavior of One-Way Concrete Slabs at Large Displacements *Applied Mechanics and Materials* **105–107** pp 1035–1039
[11] Brodzik R 2017 The influence of thermal loads on modern road concrete pavements in Poland *Scientific Journal of Silesian University of Technology Series Transport* **95** pp 27-37
[12] Lajcakova G, Melcer J 2011 Dynamic Effect of Moving Vehicles on the Road Concrete Slabs Communications *Scientific Letters of the University of Zilina* **13(3)** pp 14-18
[13] State standards of Ukraine V.2.6-98:2009 [valid from 2011-07-01] 2011 *Concrete and reinforced concrete structures. Basic provisions* Ministry of Regional Development of Ukraine p 71
[14] State standards of Ukraine B.V.2.6-2:2009 [valid from 2010-06-30] 2010 *Concrete and reinforced concrete products. General specifications* Ministry of Regional Development of Ukraine p 34
[15] State standards of Ukraine B.V.2.6-122:2010 [valid from 2011-07-01] 2011 *Reinforced concrete slabs with non-tensioned reinforcement for covering urban roads* Ministry of Regional Development of Ukraine p 23

Geometry and graphics of the fractal expansion of a square and its applications

D Tkach[1,a], **A Sopilniak**[2,b], **T Yarova**[3,c], **S Sereda**[4,d], **A Kistol**[5,e]

[1,2,3,4] Department of Descriptive geometry and graphics, Prydniprovs'ka State Academy of Civil Engineering and Architecture, 24-a Chernyshevsky St., 49600, Dnipro, Ukraine
[5]Department of Architecture, Prydniprovs'ka State Academy of Civil Engineering and Architecture, 24-a Chernyshevsky St., 49600, Dnipro, Ukraine

E-mail: [a] tkachdi@gmail.com, [b] sopilniak_artem@pgasa.dp.ua,
[c] yarova.tetyana@pgasa.dp.ua, [d] sereda.svitlana@pgasa.dp.ua,
[e] kistol.antonina@pgasa.dp.ua

Abstract. Geometry of the fractal expansion of simple geometric shapes, in particular the square, 5-celled cross and swastikas, is a new section of fractal geometry that correctly and clearlydescribes the dynamics of natural processes of growth and development of animate and inanimate cellular structures, since the iterative processes of their fractal expansion are mutuallycomplementary and inextricably interconnected. The theoretical part of this paper requires further research, and the practical one needs for the design and engineering, and technological development of a solution to the problem for building roads with the fractal-synergetic pavement asan alternative to asphalt.

Dialectics of the evolvement of modern scientific ideas about the nature of Nature and Society leads us to paradoxical concepts of chaos as a "higher form of the order, where randomness and unsystematicimpulses become an organizing principle" [1]; of fractals as "geometric shapes obtained as a result of fragmentation into parts similar to the whole ..." [2], describing such objects as coastlines, clouds, tree crowns, relief of the earth's surface, etc.; of synergetics as a science that opens up "another side of the world: its instability, nonlinearity and discoverability..." [3] and describes the properties of self- organization of both natural and social formations.

The XX century entered the history of science thanks to three fundamental discoveries that completely changed our understanding of the world: the theory of relativity, quantum mechanics and theory of chaos. The latter includes computer science, cybernetics, holography, nonlinear dynamics and fractal geometry. The achievements of these sciences have revealed the content of nonlinear processes, objects, and phenomena in nature having fractal structures and being capable of self-organization.

We are witnessing a passage from the civilization based on Aristotle's philosophy, Newton's laws and Euclid's geometry to a civilization grounded upon the philosophy of Heraclitus, laws of quantum mechanics and geometry of fractals. At the same time, the dialectical interpenetration of new scientific ideas and concepts into the well-developed traditional directions of both natural science and sciences about the artificial, i.e. manmade, is inevitable. The latter include all technical sciences and all applied areas of fundamental natural science. And since the manmade world is extremely diverse, but not always perfect, because it is being created by a person to satisfy his material and spiritual needs on the basis ofnot always perfect knowledge, it is absolutely natural for him to strive for applying new and stillunconventional knowledge to eliminate shortcomings of the old and traditional solutions. The area of architecture and civil engineering is a very wide field for such application. This raises the problem of using ideas of fractal geometry and synergetics to take the most rational engineering solutions for self-organizing designs, in particular road pavements, as well as new space-planning decisions for buildings and structures.

B. Mandelbrot's fractal geometry arose when studying the results of computer visualization of Cantor and Julia sets, Peano curves, Weierstrass functions, etc. and determined a new approach to understanding the real nature of surrounding world, whose geometry had been considered to be Euclidian since the time of Euclid and Aristotle. The author of a new geometry has proved its fractality based on Democritus's dialectical logic, according to which everything flows and everything changes, and that being invariable in the process of change has the force of law. This includes almost all animate and inanimate objects of organic origin, the kinetics of which is based on division of cells as structural elements of their constitution. In this case, before one mother cell is divided into two daughter ones, processes of doubling of all its structural elements take place in it, so that structures of daughter cells were isomorphic against the structure of the mother one. Therefore, the process of splitting into two or dichotomy is nothing more than the process of expansion or growth of the original cellular structure. Studying this dynamic process [4,5 et al.] showed that it is fractal in terms of quality, and described by Fibonacci series and metric features of golden proportioning in terms of quantity [6 et al].

B. Mandelbrot in his book "The Fractal Geometry of Nature" defines a fractal as "a structure consisting of parts, which are in some sense similar to the whole", because the infinite fragmentation and similarity of the smallest particles to the whole is a principle of the "arrangement" of nature. Various mathematicians have proposed to place abstract geometric shapes having a fractal nature, in addition to real coastlines that have infinite length in finite space, tree crowns, elements of the surface relief and clouds, the dimension of which is more than two, but less than three, etc., in a class of fractal objects. These are the Koch snowflake, Harter-Heighway dragon, Sierpiňski napkin and carpet, Cantor dust, etc. All of them are designed under the principle, which is to perform successive iterations of fragmentizing the elements of an original shape downwards the resulting self-similar shapes in their propensity for a fraction-dimensional limit.

In this regard, **a troublesome assumption** arises: if there is a real direct process of self-similar fractal reduction in the elements of an original shape, then there must be a reverse process of self-similar increase, or expansion of an original shape. If such a process exists, it becomes necessary to study and describe the dynamics of changes in shapes of successive iterations and their positional and metric properties.

Analysis of achievements and publications. In line with the logic justifying the articulation of an issue, the number of publications devoted to natural geometry and shaping in nature should include the papers of I. Sh. Shevelev [5,6 et al.], which consider the concept of the shape of an object in continuous connection with the process of its genesis. This process is dynamic as it is based on splitting of cellular structures, which causes a state of their growth or expansion. However, I. Sh. Shevelev does not reveal thefractal nature of this process, which is completely obvious both in the traditional sense of fractal structures, whose elements self-similarly decrease and by a natural assumption that these elements expectedly overgrow and expand. Such an assumption was made in the papers [7-10], where the results of iterations of the fractal expansion of a square into cross cell structures were successively obtained, and the process itself was described analytically with an output to a numerical sequence, which turned out to be the second-order Fibonacci integer sequence. The resulting geometric shapes of the 3-rd and 4-th iterations proved to be quite complex and requiring to conduct a structural analysis.

This implies setting of **a research objective:** to make a graphic-analytical description focusing on the design structure of fractal shapes of successive iterations and their combinatorial connections in order to identify an analogy with the natural process of growing cellular structures of wildlife and possible practical applications of geometry for the process of the fractal expansion of a square.

Hard core of the paper. If we take the most technological square as an original shape and subject it to a sequential iterative expansion by adding the results of its previous transformations to the subsequent results of iterations, then shapes of the fractal nature will appear. As a result of the first iteration, thesquare expands or transforms conformally and topologically into a 5-celled cross; as a result of the second iteration, the central square of the 5-celled cross expands or similarly transforms into a 9-celled square, and its 4 ends expand up to the 5-celled crosses. In sum, there is a 29-celled shape inscribed in an 81- celled square with 4 axes of symmetry. Its structure includes the results of the 1-st iteration in the form of four 5-celled crosses.

As a result of the third iteration, the 29-celled shape undergoes the topological transformation into a 169-celled shape. Upon that, the central 9-celled square turns into a 25-celled central square-lack, and the 5-celled crosses of the 2-nd iteration in the middle of its sides turn into the 29-celled shapes.

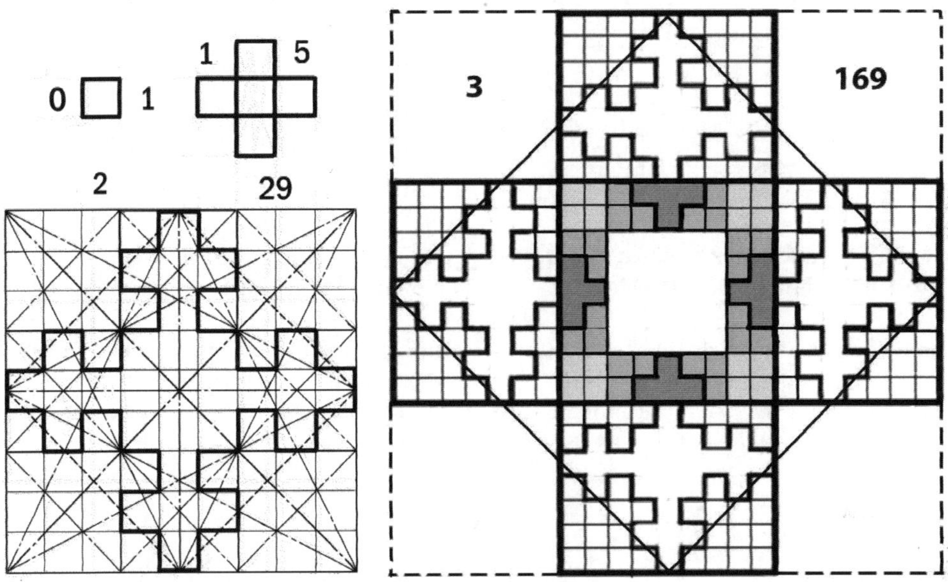

Fig.1. Three iterations of the fractal expansion of the square

In doing so, the overall 81-celled squares superimpose with their 18 cells on the central 81-celled square, and each 29-celled shape seems to "sink" in the central square with its 4 cells.

IOP Conf. Series: Materials Science and Engineering 1162 (2021) 012018 doi:10.1088/1757-899X/1162/1/012018

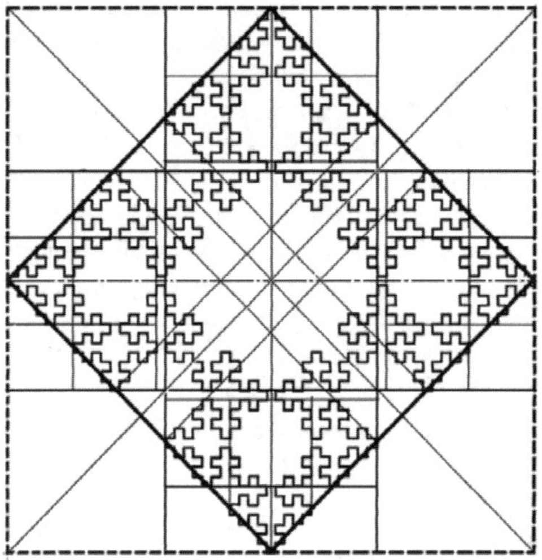

Fig.2. Shape of the 4-th iteration of the fractal expansion of the square

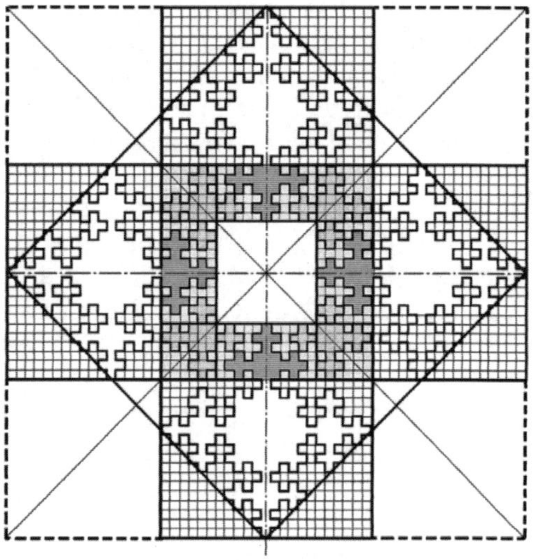

Fig.3. "Intergrowth" of lateral shapes of the 4-th iteration of the square expansion with the central square

The fourth iteration topologically and conformally transforms the 169-celled shape into a 985-celled one, the structure of which includes four 169-celled shapes, whose dimensions self-similarly repeat the 5- celled cross and "sink" with their 4 twenty-fourth parts in the central square (Fig.3). In terms of design,the process of this "intergrowth" occurs through the appropriate structuring of shapes of the overall square's superimposition.

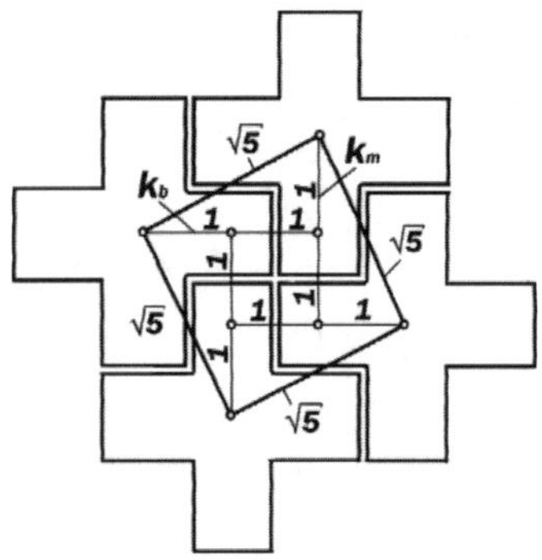

Fig.4. Cross packing

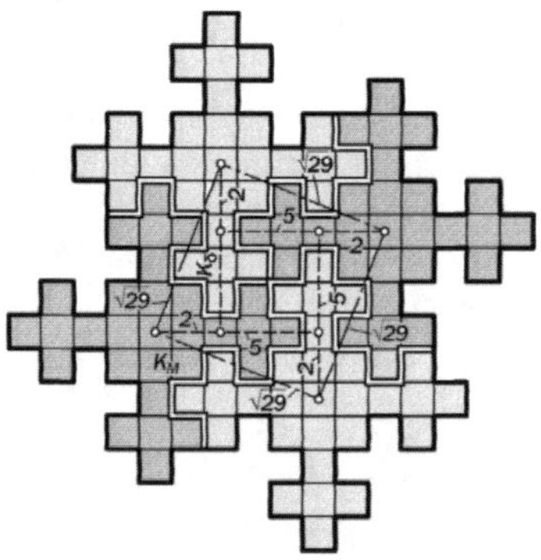

Fig.5. Packing of the 29-celled shapes

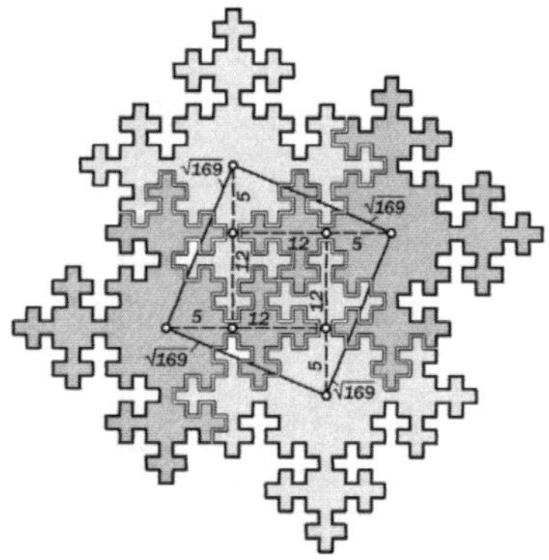

Fig.6. Packing of the 169-celled shapes

It should be especially noted that the shapes of the considered iterations are rooted in the 5-celled cross of the overall squares, inside which the "cell division" takes place. If we assume that the sizes of these squares are unchanged, then increasing the number of cells in each of them leads them to the self-similar reduction, which corresponds to the traditional assumption of the fractal nature of natural systems.However, if we consider the sizes of the original square as unchanged, the process of increasing the number of cells congruent to it in the shapes of subsequent iterations causes its fractal expansion within the accordingly increasing overall squares of the 5-celled cross. In both cases, as the shapes of subsequent iterations become more complex, the structures of the lateral squares make a union with the central one, or there is a process of their "growing" from the central square. Likewise, the dormant buds of a plant generate the first-order branches; overtime, their buds generate the second-order branches and so on. The same processes are observed in living organisms.

The results of successive iterations of the original square cell can be represented by elements of combinatorial compositions that densely pack the plane.

The 5-celled crosses compactly pack the plane being adjacent to each other at the swastika-shaped joints (Fig.4). As a quantitative characteristic of their relative position, one can take the ratio of the legs (1:2) of right-angled triangles, the hypotenuses of which connect their centers. The sum of their squares (1+4) is equal to 5, i.e. the number of cells of one element.

The combinatorics of four 29-celled elements compactly packs the plane owing to their entry into the lock joints (Fig.5). The ratio of the legs of triangles, the hypotenuses of which connect their centers, is equal to 2:5. The sum of their squares (4+25) is equal to 29, i.e. the number of cells in one element (Fig.5).

The combinatorics of combining four 169-celled elements into the lock forms their system as the integrated whole. The straight lines that connect their centers are the hypotenuses of right-angled triangles, the lengths of whose legs are related as 5/12. The sum of their squares (25+144) is equal to the square of the hypotenuse (169), i. e. the number of cells in one element.

Following the accepted technique of transforming previous iterations into subsequent ones, it is possible to obtain shapes becoming more and more complex in their geometric structure and expanding up to infinitely larger sizes.

Among the distinctive features of dense packings of the plane with various fractal shapes, there is their synergism or ability to self-organization, since the lock joints between them redistribute stresses from possible loads and force the prefabricated design, for example that of a roadway, to operate as monolithic one but crack-resistant, inasmuch as the role of cracks is played by joints between the elements.

A comparative assessment of the process to change values of the leg ratio in right-angled triangles, the hypotenuses of which connect the centers of these shapes, allows determining a quantitative law of the entire iterative process (Table 1).

Table 1. Quantitative characteristics of the iterative process towards the fractal expansion of the square

Iteration No. (n)	0	I	II	III	IV	V	VI
K_b – long leg	**1**	**2**	**5**	**12**	**29**	**70**	**169**
K_m – short leg	**0**	**1**	**2**	**5**	**12**	**29**	**70**
Number of cells	**1**	**5**	**29**	**169**	**985**	**5741**	**33461**
Number of sides	**4**	**12**	**52**	**220**	**932**	**3948**	**16724**

An analysis of the above Table indicates the following pattern:

$$K_{b(n)} = 2\,K_{b(n-1)} + K_{m(n-1)}.$$

Since $K_{m(n-1)} = K_{b(n-2)}$, then $K_{b(n)} = 2\,K_{b(n-1)} + K_{b(n-2)}$,

whence a new numerical series follows: $a_n = 2\,a_{(n-1)} + a_{(n-2)}$ (1)

This expression is a recursion formula reflecting the iterative process of expansion of the square.

The numerical series of changing the number of cells in the shapes starting from the 2-nd iteration is described by the expression: $a_n = 6\,a_{(n-1)} - a_{(n-2)}$ (2)

The peculiarities of changes in the number of sides of the sequential results of iterations, starting from the 2-nd one, are described by the expression below:

$$a_n = 4\, a_{(n-1)} + a_{(n-2)} \quad (3)$$

A special attention should be paid to the fact that the validity of the expressions (2) and (3) begins from the second iteration. This is explained by emergence of the process, when its regular patterns described using these expressions appear in discrete steps and do not cover the description of its beginning.

Considering the positional properties of dense packings, we draw attention to the swastika-shaped joints, as it were, clamped between the cross elements. If one mentally imagines that they expand by means of corresponding crosses up to the equality of their width, then one gets a two-element packing of crosses and swastikas, which self-closes by virtue of emerging synergy.

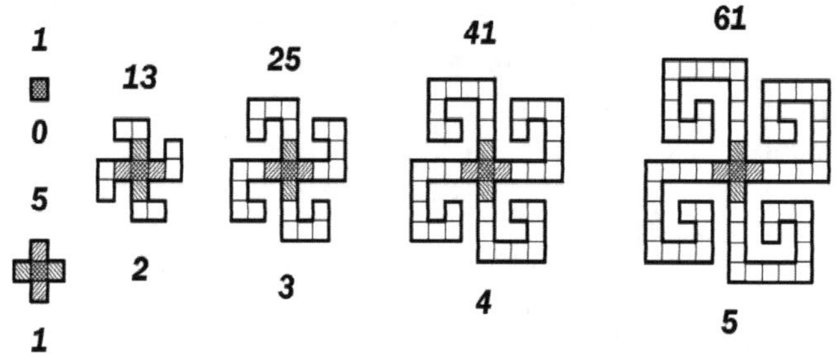

Fig. 7. 5 iterations of the fractal transformation of the cross into swastikas

If the process of compressing crosses continue downwards the level of the width of a jointbetween the elements, then a swastika structure appears, theelements of which can also be represented as the results of therelevant fractal expansion of thesquare (Fig.7).

As a result of the first iteration, the original square transforms into a 5-celled cross, and as a result of the second iteration, following the swastika structure, double squares join the lateral cells of the cross, so that a 13-celled swastika appears.

IOP Conf. Series: Materials Science and Engineering 1162 (2021) 012018 doi:10.1088/1757-899X/1162/1/012018

The third iteration complements the swastika cross base by 5 П-shaped cells and a 25-celled swastika is obtained.

The fourth iteration complements the cross base by 9 specially arranged cells and a 41-celled meander swastika is obtained and so on.

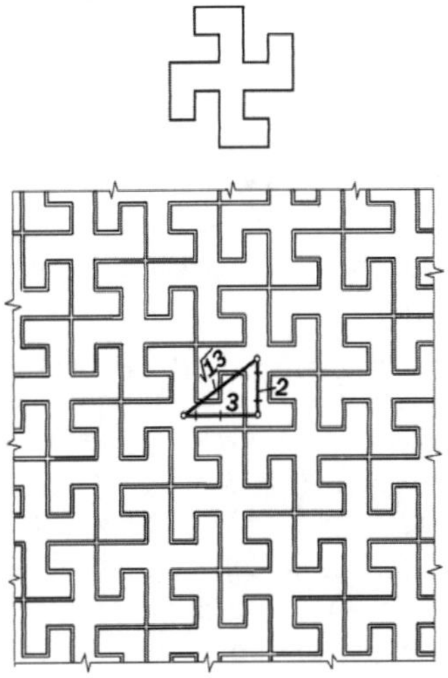

Fig.8. Packing of the 13-celled swastikas

Fig.9. Packing of the 25-celled swastikas

In practice, the original square, having transformed into the 5-celled cross, gives it the properties of an initiator of all subsequent iterations.

It is not hard to see that the swastika shapes of the 2-nd and 3-rd iterations tightly pack the plane (Fig.8 and 9), as well as to assume that the shapes of the 4-th, 5-th and subsequent iterations of the 5-celled cross also pack the plane compactly, and their metric characteristics easily extrapolate on the basis of available information about sizes of the legs of right-angled triangles, the hypotenuses of which connect the centers of these shapes (Table 2).

Table 2. Quantitative characteristics of the iterative process towards the fractal expansion of the 5-celled cross

Iteration No. (n)	0	I	II	III	IV	V	VI	VII
K_b – long leg	**1**	**2**	**3**	**4**	**5**	**6**	**7**	**8**
K_m – short leg	**0**	**1**	**2**	**3**	**4**	**5**	**6**	**7**
Number of cells	**1**	**5**	**13**	**25**	**41**	**61**	**85**	**113**

A statistical analysis of this Table shows that:

$$K_{b(n)} = 2K_{b(n-1)} - K_{m(n-1)}$$

Since $K_{m(n-1)} = K_{b(n-2)}$, *then* $K_{b(n)} = 2K_{b(n-1)} - K_{b(n-2)}$ (4)

It turns out that the expression (4) describes a series of natural numbers:

$$a_{(n)} = 2a_{(n-1)} - a_{(n-2)} \quad (5)$$

or **- n, … -8,-7,-6,-5,-4,-3, -2, -1, 0, 1, 2, 3, 4, 5, 6, 7, 8,… n**

As a result, it turns out that the expression (5) describes an integer sequence (3). But it is fundamentally different from the generally accepted formula of the integer sequence $a_{(n)} = a_{(n-1)} + 1$ n thatit follows from the quantitative analysis of features revealing the iterative process of fractal transformations of the 5-celled cross, which spatially models it. This means that any operations with numbers indirectly induce the corresponding operations with structures of the fractal expansion of swastikas.

The comparison of the expressions (1) and (5) points to the similarity of their elements in the right side and their difference in signs between these elements. This circumstance provides evidence of not the opposite, but the mutual complementarity and inextricable connection of processes of the fractal expansions of the square and 5-celled cross into swastikas.

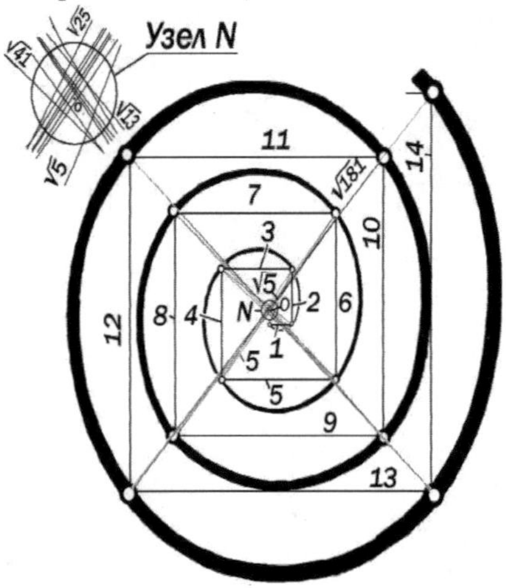

Based on the analysis of Figures 2 and 3, and Table 1, it follows that the longer legs of previous iterations act as the shorter legs of subsequent iterations. Therefore, it is easy to imagine the dynamics of development of the synergetic swastika structures in the form of a spiral by applying the longer leg of the subsequent iteration to the end of the previous longer leg at a right angle (Fig.10).

In this case, the ratios (2:1=2; 3:2=1,5; 4:3=1,33; 5:4 =1.25; 6:5=1,2; 8:7 =1,14; etc.) of the leg lengths intriangles of subsequent iterations are not the same, and therefore, their hypotenuses do not intersect at one point, but limit in the picture a certain region (node N), which, after each iteration, decreases in size singulating to some point *O* lying outside the construction triangles.

Fig.10. Visualization of the integer sequence

Thus, we can assume that this region "shuts" in the point *O* at the number of iterations $n = \infty$, which will divide the infinitely large hypotenuse in half. This can happen when the legs of the infinitely large triangle will become equal. It follows that the development of the spiral in Fig.10 shows the evolution of passing from the Dürer triangle (hypotenuse $\sqrt{5}$) to a right isosceles triangle.

Since fractal geometry basically visualizes natural linear and nonlinear processes in nature rather peculiarly, the proposed geometry of the fractal expansion of simple geometric shapes also models, on a geometrical and graphical basis, a process of growth of living cellular structures in the form of generating tree crowns and behaviors of climbing plants, in particular vine whiskers. Although, these models are more specific, since they are more logical and visual.

Applications: 1. One-layer synergetic pavement of the earth's surface.

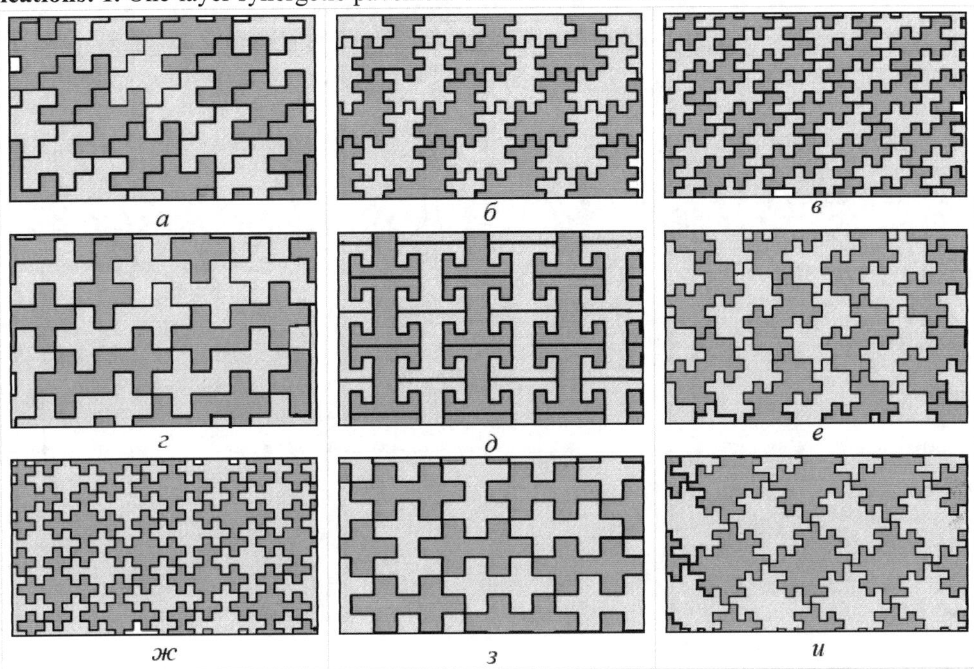

Fig.11. Options of one-layer synergetic pavements for footpaths

It is carried out by the effect from lateral shapes of various iterations of the fractal expansion of the square being intergrown with the central square of the overall 5-celled cross, leading to the entry of congruent shapes into the lock joints of dense packings of the plane, which forms the concept of a one- layer prefabricated synergetic pavement for footpaths (Fig.11) and large areas as an alternative to their asphalt pavement. The idea of a prefabricated pavement is widely popular in European and other countriesof the world for paving footpaths, small areas, parking lots and adjacent territories. Nevertheless, as in Ukraine, in the world such paving does not have a synergetic effect of self-organization, since its tiles do not enter into the lock joints between themselves. Moreover, one should expect from this effect savings in the material of tiles due to redistributing loads between them, reducing the construction timeframe and increasing the aesthetic appeal of the entire pavement.

2. Two-layer synergetic pavement of the carriageway.

It is carried out using the positional feature, inherent in the shapes of the 3-rd and 4-th iterations just to name a few (see Fig. 1 – 3), to be located in overall squares, the vertices of which divide in half the sides of the overall square of their "5-celled cross". A composition of "a square in a square" is obtained, when the area of the inscribed square is equal to the total area of 4 angled triangles of the overall square of the cross. This circumstance led to the idea of making a square recess in a solid square element by half its thickness that gives rise, when 4 such elements are relatively positioned, to the formation of a protruding square composed of 4 angled triangles, the area of which will be equal to the area of a recessed square. If the areas are calculated taking into account the width of joints between the elements, then the element of the upper layer with its square recess is "slipped" over the "square" prefabricated of 4 angled triangles, and thereby, "closes" 4 elements of the lower layer into a "lock".

IOP Conf. Series: Materials Science and Engineering 1162 (2021) 012018 doi:10.1088/1757-899X/1162/1/012018

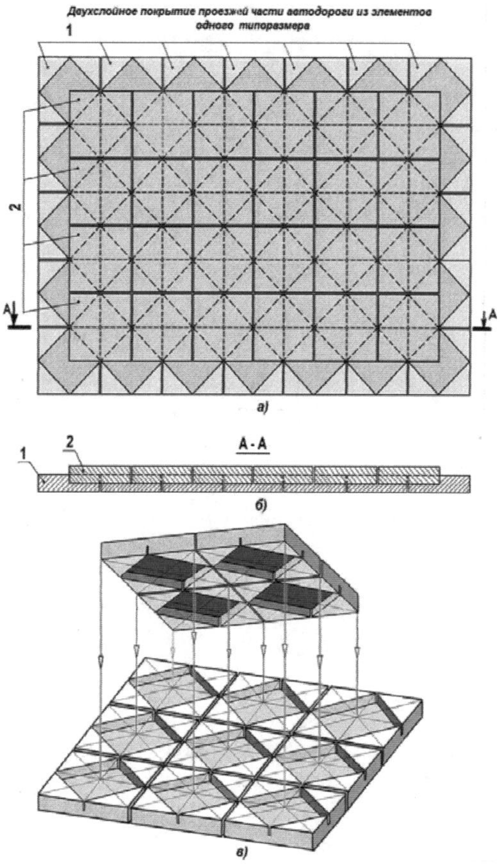

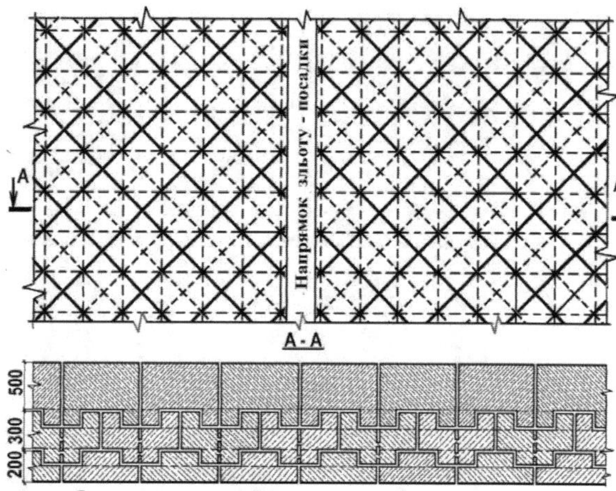

Three-layer synergetic pavement of the airport runway

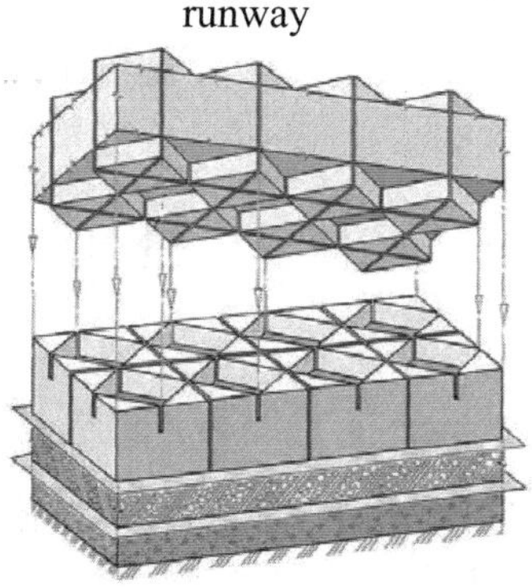

Fig.12. Two-layer synergetic pavement of the carriageway

Fig.13. The second and third pavement layers

As a result, we get a two-layer prefabricated synergetic pavement of the carriageway, which operates as a monolithic but crack-resistant one, since the role of cracks is played by joints between the elements.

3. Three-layer synergetic pavement of the airport runway.

The airport runway is a specially prepared and equipped landing strip with an artificial pavement, which ensures takeoff and landing of an aircraft. It is traditionally a multi-layer design. In recent years, steel fiber concrete has been used as the material of the upper layer, being heavy-weight concrete with appropriate additives to resist the effects of high and low temperatures, reinforced with steel fiber, i.e. a mixture of steel fibers, which should be uniformly distributed in bulk concrete.

The high labor intensity of construction works to build the airport runways determines accordingly their high prime cost, and therefore, is a reason to search for alternative solutions to the problem of their building as durable, reliable and competitive facilities.

A worthwhile basis for such a search is the above-described concept of the carriageway synergetic pavement, which consists of the elements of industrial production being congruent in each layer, that is, identical both in shape and in size, which enter into the lock joints between themselves and thereby force all prefabricated pavements to be hard at work as the monolithic (Fig.13).

4. One-layer synergetic pavement of the water surface from congruent pneumatic elements

The positional property of the flat 29-celled fractal shapes (Fig.14, a) to enter into the permanent lock joints served as the basis for the idea of giving them a third dimension in the form of 52-faced surfaces made of high-duty membrane (Fig.14, b).

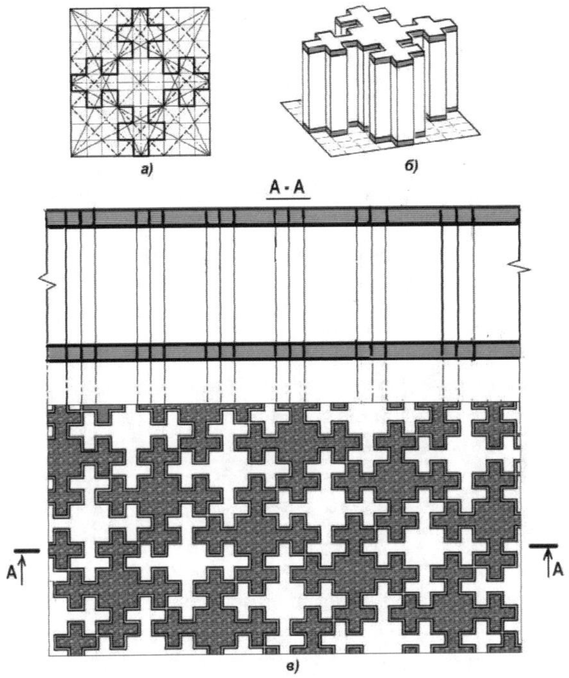

Fig.14. Pavement of the water surface

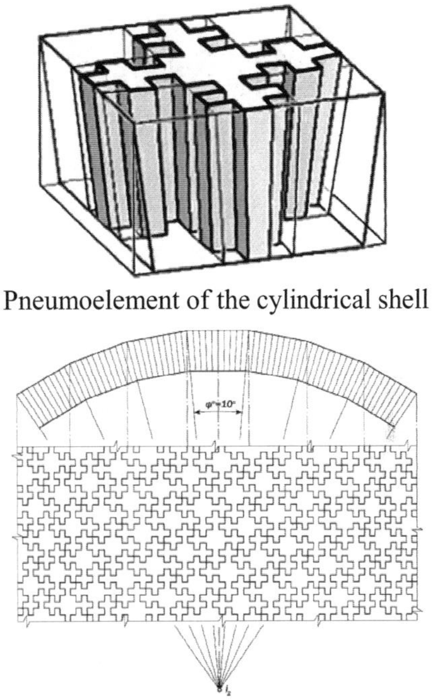

Fig.15. Cylindrical pneumoshell

If the volumetric interpretation of this shape is a 52-faced prism, then a flat layer of space, sort of roadway paving, will be structured with these shapes. Being extremely light, it can spread on water and serves as a pontoon bridge, or an artificial island.

If the faces, which are symmetric to one of the two planes of the 52-faced prism symmetry, are given a design-level slope, it is possible to mount a cylindrical shell of any span from the obtained wedge-shaped elements. It seems that due to lightness of such shells and strength of links between them, they can overlap the large areas and entire structures, and serve as large-span hangars, warehouses and other objects for various purposes.

If we assemble from them a close cylindrical shell, capable of aeronautics, in case of its elements being filled with light gas, then we can talk about the creation of reliable airships and other aircrafts (Fig.15).

If the 52-faced prism is transformed into a truncated pyramid, then a hemispherical or spherical shell of the large span can be assembled from these elements as an independent building or air balloon (Fig. 16, 17).

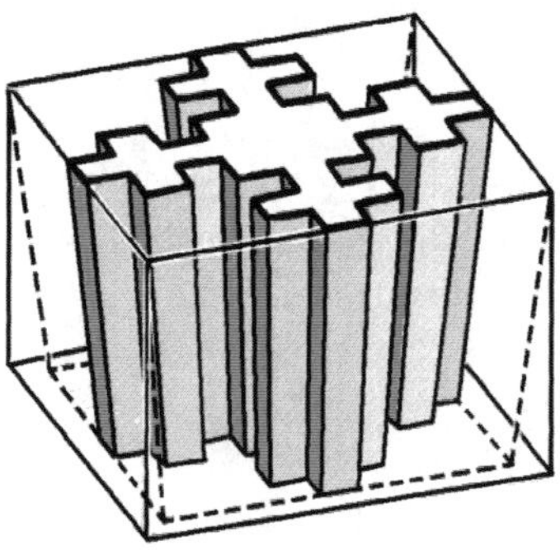

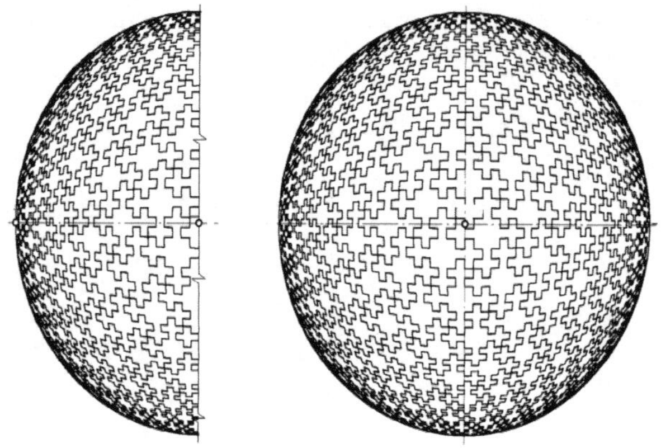

Fig. 16. Element of the fractal sphere **Fig. 17**. Hemispherical fractal shell

The advantages of such shells are the absence of necessity to maintain excess pressure inside the structure, their large-span and durability. Failure of one or more elements does not lead to destruction of the design, but only requires their replacement. Only the design of the air supply valves is troublesome.

Conclusions

1. Geometry of the fractal expansion of simple geometric shapes, in particular the square, 5-celled cross and swastikas, is a new section of fractal geometry that correctly and clearly describes the dynamics of natural processes of growth and development of animate and inanimate cellular structures, since the iterative processes of their fractal expansion are mutually complementary and inextricably interconnected.
2. A wide range of possible fundamentally new practical applications of geometry of the fractal expansion of the square, based on the unrestrained correspondence to their nature, determines the real opportunity of creating easy-to-manufacture, economically profitable, and competitive material objects, such as dense packings of the earth's surface of the type of carriageways, footpaths, airport runways, earthwork structures, floating water crossing installations, artificial islands and marine terminals, large-span arch anddome buildings, self-supporting fences in a supporting framework of high-rise objects, aerostats, etc.
3. The theoretical part of this paper requires further research, and the practical one needs for the design and engineering, and technological development of a solution to the problem for building roads with the fractal-synergetic pavement as an alternative to asphalt.

References

[1] Tkach D, Nifanin A 2014 From Chaos to Order *LAMBERT Academic Publishing, Saarbrucken, Deutschland* p 98
[2] Kniazeva Y, Kurdiukov S 2005 The Foundation of Synergetics. Synergetic Worldview *MoscowURSS* p 165
[3] Tkach D 2006 Geometry of Crack-Resistant and Self-Closing Structures of Road Pavements *System technologies* **3(44)** p 121-127
[4] Nifanin A, Tkach D 2006 Geometry of Two-Layer Synergetic Structures and Its Applications *Bulletin of Kyiv National University of Technologies and Design* **4** p 90-95
[5] Tkach D, Nifanin A 2007 The Gold Ring Geometry and Its Applications *Geometric and Computer*

Modeling **18** p 147 – 153
[6] Tkach D, Kistol A, Nifanin A 2009 The Fractal Expansion of a Square as a Geometric and Graphic Model of the Process of Growth of Wildlife Cellular Structures *Geometric and Computer Modeling. Collection of Research Papers* **25** з 170-175
[7] Nifanin A, Tkach D 2009 The Group of Golden Irrationalities as a Metric Basis of Geometry of the Fractal Expansion of a Square Geometric Modeling and Computer Technologies *Theory, Practice, Education* p 180-191
[8] Tkach D, Kistol A, Nifanin A 2007 Geometry of the Fractal Expansion of a Square and Its Applications *Civil Engineering, Materials Science, Mechanical Engineering* **41(3)** p 118-123
[9] Shevelev I 1986 The Principle of Proportion *Moscow Stroyizdat* p 345

IOP Conf. Series: Materials Science and Engineering 1162 (2021) 012019 doi:10.1088/1757-899X/1162/1/012019

Crack formation and measurement in concrete with carbon fiber-reinforced composites under bending test

Yuriy Vashpanov [2], Jong Gun Park[1], Tatyana Podousova[3], Gwang Hee Heo[1], Jung Young Son[1]

[1]Department of Civil Engineering, Public Safety Research Center, Konyang University, Nonsan, Chungnam, 32992, Republic of Korea
[2]Department of Physics, [3]Department of Applied Mathematics, Odessa State Academy of Civil Engineering and Architecture, Odessa, 65029, Ukraine

the corresponding author's e-mail address yuriy.vashpanov@gmail.com

Abstract. The certified test equipment of the MTDI Inc. Materials Testing System under bending test was using for cracking in concrete with carbon fiber-reinforced composites. The concrete samples included microfibers 6 and 12 mm long with a total volume fraction of fibers from 0.5 to 2.0%. To record the results of the experiment, a high-speed video camera was used, since with an increase in the value of the load, cracking and then destruction of the sample can occur at any moment, which is unpredictable when using different samples. This significantly increases the accuracy of determining the moment of crack formation and describing the geometric dimensions of cracks depending on the load. Optical non-contact methods of the measurements and image processing is used for determining the geometric dimensions of cracks in specimen. For concrete containing carbon fibers 6 mm long, with an increase in load and volume fractions ν cracking occurs at loads that are 1.67 times lower than in specimens with 12 mm filaments. Concretes with 12 mm long carbon fibers had a higher time for crack formations (the load increased linearly with time on Bending jig of the Materials Testing System), which slightly increases by 1.13 times with increasing parameter ν from 0.5 to 2.0%. Cracks in concrete in both cases are observed at practically the same load magnitudes at values of ν =2%.

Introduction

Optical non-contact methods of diagnostics and cracks monitoring of building structures are currently being actively studied, as evidenced by numerous publications over the last year [1-12]. The proposed new optical non-destructive methods accelerate and significantly reduce the cost of monitoring the state of concrete structures, which is undoubtedly useful for the development of the construction industry. The creation of new, more durable building materials designed for increased loads has always been relevant. Currently, fibers used in fiber-reinforced cement composites include deferent fibers such as steel, glass, carbon, aramid, polypropylene, and nylon etc. One of the main problems in the operation of building structures is the appearance of cracks in them, which subsequently leads to their destruction. Fiber-reinforced cement composites are manufactured by incorporating fibers and can suppress the opening and propagation of cracks due to the bridging action of fibers [13]. Currently, it is important to develop the most accurate definition of the geometric parameters of cracks in concrete, depending on the technology of concrete production and ultimate loads. This research is based on the results of our paper for determining the geometric dimensions of cracks in concrete be optical method published in [14]. In this paper, the processes of cracking of concrete depending on the cement-fiber-cement composites technology and optical measurements cracks geometrical parameters depending on load values and type of concrete composites is discussed.

Experimental

Figure 1, a) shows the certified test equipment of the MTDI Inc. Materials Testing System (Republic of Korea) for experiments. The concrete material testing system is fully automated with a computer and contains a load cell (sensor) with an automatic calibration function. The sensor is connected via a controller to accurately measure and control the load. The computer can programmatically set the rate of increase of the load with high accuracy and repeatability. The MTDI Inc. Materials Testing System can be used for many different types of concrete testing tasks and is fully computer controlled. The crosspiece has an excellent stiffness of 287 kN/mm, which is better than other manufacturers.

During bending tests at various significant loads, cracks appear on samples from different grades of concrete. To carry out such tests, the samples were placed on special supports 2 (Figure 1, b). The load increased linearly with time on Bending jig 1 (Figure 1, b). The rate of increase in the load was set by the control computer. A high-speed video camera was used to record the results of the experiment, since with an increase in the load, cracking can occur and then the destruction of the sample, which can be unpredictable when using different samples. The camera was synchronized with the control computer. This allows, when analysing video materials, to record the moment of crack formation in the sample as accurately as possible, and then to measure as accurately as possible the geometric parameters of the crack with increasing load (different times from the beginning of the load growth) using non-contact optical measurement method [14].

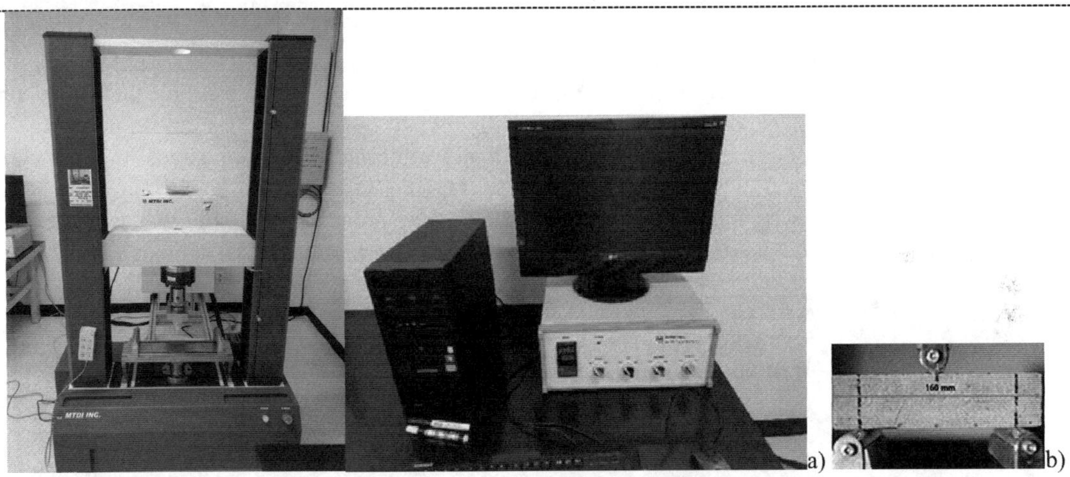

Figure 1. a) Certified test equipment of the MTDI Inc. Materials Testing System (Republic of Korea), b) a specimen of concrete for bending test experiment.

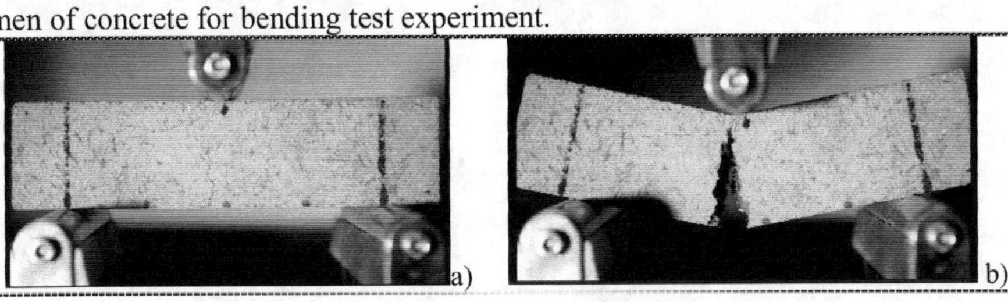

Figure 2. a) The photo at moment of time found from the analysis of the video data of the high-speed camera, corresponding to the formation of a crack with increasing load. b) Sample completely destroyed.

Figure 2, a) shows a snapshot of a sample at a time with an already noticeable crack in a sample of concrete. An increase in the load leads to the destruction of the sample (Figure 2, b). This bending test provides data on the maximum load (crack formation in a concrete structure) for various concrete production technologies. Crack formation in concrete with carbon fiber-reinforced composites under bending test using the Materials Testing System were compared and analysed.

Results and discussion

Carbon fibers had different lengths 6 mm and 12 mm. The concrete with carbon fiber-reinforced composites had a mixture with fibers corresponding to volume fractions v 0.5%, 1.0%, 1.5%, and 2.0%. Figure 3 shows photographs of the formation of cracks in the test samples, which are already clearly visible under the conditions of this experiment with high-speed video recording with the start time of the process of increasing the load. The bending load of samples of different manufacturing technologies is proportional to the time. Therefore, a relative comparison of the experimental data is performed using the time parameter t.

Cracking is individual for each specimen, but cracks appear near the bending jig, where the load is maximum in magnitude (Fig. 3). The concrete with carbon fibers is formed technologically by chaotic inclusions of components of cement, sand and fibers additives. Consequently, a crack is formed when the weakest link in the concrete structure breaks. It is known that concrete can contain inclusions of microvoids filled with air [15]. The presence of microvoids is determined by a special test for weighing ready-made concrete and its components. The difference in the values of the weights signals the presence of such voids. Such microvoids form ruptures of interatomic bonds and deteriorate the strength of concrete. In this case, microcavities can act as areas with low microhardness and a weak link in the concrete structure that is resistant to loads. These voids can act as a crack growth generator since they are microcracks themselves. With an increase in the load, the areas near microvoids (microcracks) will expand to form visually visible macrocracks, the shape of which is practically unrepeatable (Fig. 3) and depends on the biography of the creation of a particular sample. The second reason for the formation of cracks is the presence of inhomogeneous inclusions with lower intermolecular forces of interaction than in the rest of the concrete structure. Finally, at high loads, any hardest homogeneous material can be destroyed, which is associated with its ultimate strength. The creation of new types of concretes that can withstand significant loads for a long time without cracking is the most important technological task at the present time.

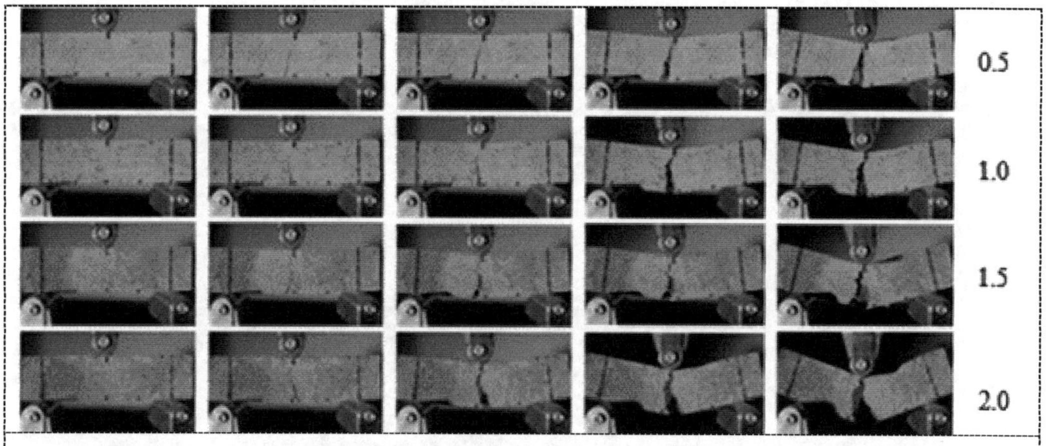

Figure 3. The photos of a crack formation for different samples under bending test experiment.

Figure 4 (a) shows maximal crack widths in concrete samples with fibers 6 mm length corresponding to volume fractions v 0.5%, 1.0%, 1.5%, and 2.0% from working time of the MTDI Inc. Materials Testing System under bending test. Experimental dependences of fracturing-formation time for different samples with different lengths of fibers with 6 and 12 mm and concretes with fibers mixture corresponding to volume fractions ν is shown on Figure 4 (b).For concrete containing carbon fibers 6 mm long, with an increase in load and parameter ν cracking occurs at loads that are 1.67 times lower than in specimens with 12 mm filaments. Concretes with carbon fibers 12 mm long have a higher cracking time (higher loads), which slightly increases by 1.13 times with an increase in the parameter ν (Figure 4 (b)). Cracks in concrete in both cases are observed at practically the same load magnitudes at values of ν =2% (Figure 4, b).

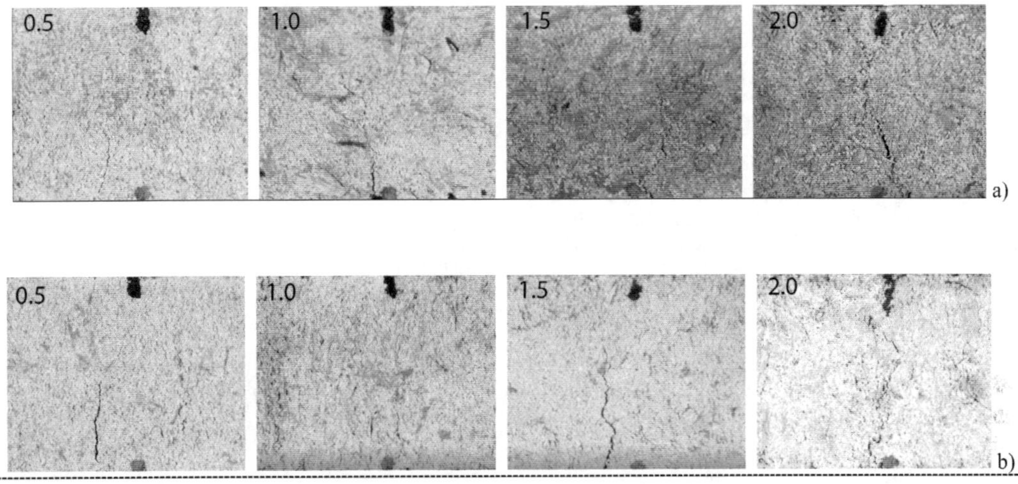

Figure 4. Photos for visible cracking moments from video recording data by high-speed camera for different samples with different lengths 6 mm (a) and 12 mm (b) and concrete have a mixture with fibers corresponding to volume fractions ν 0.5%, 1.0%, 1.5%, and 2.0%.

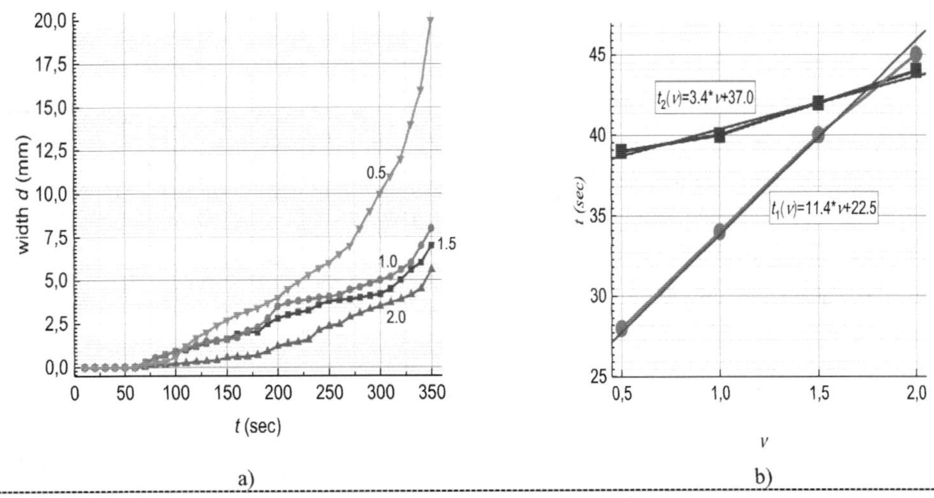

Figure 5. (a) maximal crack widths in concrete samples with fibers 6 mm length corresponding to volume fractions ν 0.5%, 1.0%, 1.5%, and 2.0%, (b) dependences of fracturing-formation time for different samples with different lengths (6 mm (red) and 12 mm (black)) and concrete have a mixture with fibers corresponding to volume fractions ν.

Figure 5 (a) shows maximal crack widths in concrete samples with fibers 6 mm length corresponding to volume fractions ν 0.5%, 1.0%, 1.5%, and 2.0%. Figure 5 (a) shows (b) dependences of fracturing-formation time for different samples with different lengths (6 mm (red) and 12 mm (black)) and concrete

IOP Conf. Series: Materials Science and Engineering 1162 (2021) 012019 doi:10.1088/1757-899X/1162/1/012019

have a mixture with fibers corresponding to volume fractions ν. Figure 6, a) and b) shows a linear approximation of the experimental data on the appearance of a crack in concrete specimens with different fiber lengths 6 and 12 mm and at different value of parameter ν with the found values of statistical errors.

Linear fitting curves for experimental data $t(\nu) = a + b \cdot \nu$ is be received by mathematical module of the Origin Lab program. The fitting curves for samples can be present

$$t_1(\nu) = a_1 + b_1 \cdot \nu, \quad a_1 = 37.0 \pm 0{,}474; b_1 = 3{,}4 \pm 0{,}3464 \tag{1}$$

$$t_2(\nu) = a_2 + b_2 \cdot \nu, \quad a_2 = 22{,}5 \pm 0{,}474; b_2 = 11{,}4 \pm 0{,}346 \tag{2}$$

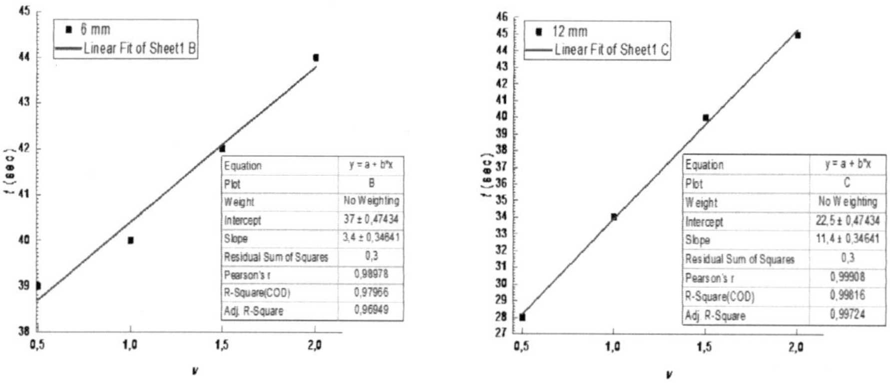

a) b)

Figure 6. A linear approximation of the experimental data on the appearance of a crack in concrete specimens with different fiber lengths 6 and 12 mm and at different value of parameter ν with the found values of statistical errors.

The experimental data considered above testify to the processes of inhibition of the formation of cracks in concrete when filamentary carbon materials are included. Figure 7 shows a physical model for the role of carbon fiber in inhibiting concrete cracking in flexural tests. Carbon fibers are associated with concrete by mechanical stirring and the directions into the structure of the material must be chaotic (Fig. 7). Carbon fibers can hinder crack growth due to the elastic forces of the fibers, which are in the region of maximum load when the interatomic bonds in the concrete are already broken. In the initial stages of cracking with increasing load, the elongation of the carbon fiber restrains the rapid destruction of concrete. The crack can disappear due to the restoration of the original unloaded length of the fibers during unloading. The displacement of carbon fibers from the places of fixing-in in the bulk of concrete are more likely at significant loads (indicated by white rectangles (voids) (Figure 7)). Irreversible changes in the structure and shape of the concrete structure will be observed. Longer carbon fibers have more surface contact with surrounding concrete components and therefore have a better fixing-in with concrete. Carbon fiber perpendicular to the crack direction (parallel to the sample surface, Figure 7) has a better fixing-in with concrete and, therefore, should inhibit crack growth to a bigger extent. Thus, the parallel arrangement of carbon fibers with respect to the surface (perpendicular to the load force) can lead to a decrease in cracking in concrete at significant stresses. However, it will be difficult to implement technologically at the present time.

Thus, the physical reason for the increase in load by a factor of 1.67 for the formation of a crack with longer fibers (length 12 mm) with volume fractions ν = 0.5% may be a better fixing-in of long carbon fibers in concrete.

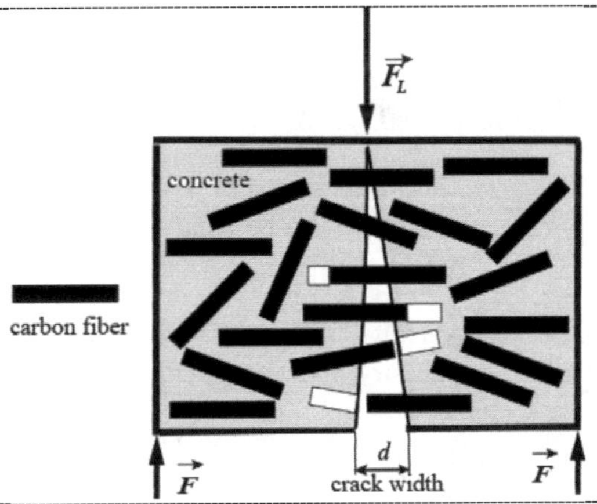

Figure 7. A crack formation model in concrete with carbon fiber-reinforced composites under bending test.

Conclusion

This study presents a novel optical method approach for research of crack formation and its measurement in concrete with carbon fiber-reinforced composites under bending test. For concrete containing carbon fibers 6 mm long, with an increase in load and parameter ν cracking occurs at loads that are 1.67 times lower than in specimens with 12 mm fibers. The concrete with carbon fibers of the length 12 mm have a higher fracturing-formation time, which increases slightly by 1.13 times with an increase in the parameter ν. Cracks in concrete in both cases are observed at practically the same load magnitudes at values of ν=2%.

References

[1] Miao Zenghui, Ji Xiaodong, Okazaki Taichiro, 2021*Computer-Aided Civil and Infrastructure Engineering,* 1-18.

[2] Chuan-Zhi Dong and F Necati Catbas 2020 *Structural Health Monitoring* 1-52.

[3] Mei Q and GuË l M. 2020 *Construct. Build. Mater.* 256, 119397.

[4] Gehri N., Mata-Falcón J., Kaufmann, W. 2020 *Construct. Build. Mater.* 256, 119383.

[5] Ji X, Miao Z and Kromanis R. 2020 *Eng Struct* 212, 110508.

[6] Al-Mistarehi Bara' Wasfi, Obaidat Mohammed Taleb 2020 *International Journal of Advanced Science and Technology*, 29, No. 4s, 2465 - 2480.

[7] Rui L., Zappa E., Collina A. 2020 *Eng. Fract. Mech.* 224, 106715.

[8] Słonski Marek and Tekieli Marcin 2020 *Materials* 13, 3527.

[9] Song Weidong, Jia Guohui, Zhu Hong, Di Jia, and Lin Gao 2020 *Journal of Advanced Transportation*, Article ID 6412562, 11 pages.

[10] Feng Chuncheng, Zhang Hua, Wang Haoran, Wang Shuang and Li Yonglong 2020 *Sensors* 20, 2069.

[11] Shim Seungbo, Kim Jin, Cho Gye-Chun, and Lee Seong-Won 2020 *IEEE Access* 8, 170940-170950.

[12] Won Kwanghee and Sim Chungwook 2020 *Sensors*, 20, 1838.

[13] Heo G.H, Park J.G, Song K.C, Park J.H, Jun H.M. 2020 *Advances in Civil Engineering*, 8881273.

[14] Vashpanov Yu., Son J.Y., Heo G.H., Podousova T., and Kim Y.S. 2019 *Advances in Civil Engineering*, 2398124.

[15] Becker H.H., Ley M.T., Welchel D., Peery J., Leflore J., Brorsen B.W., Khatibmasjedi M. 2021 *Construction and Building Materials*, 288, 122865.

Models of co-evolution of structural elements of building composites

V Vyrovoy [1,a], V Sukhanov [1,b], A Ielkin [2,c] and S Sukhanova [1,d]

[1] Department of Production of Building Products and Constructions, Odessa State Academy of Civil Engineering and Architecture, Staroportofrankovskaya st., 28, 65020, Odessa, Ukraine

[2] Construction Company «BUDOVA», Osipova st., 25, 65012, Odessa, Ukraine

E-mail: [a] vyrovoy@ukr.net, [b] ecostroy_odessa@ukr.net, [c] ielkin@gmail.com, [d] svetlanasukhanova@rambler.ru

Abstract. It is shown that construction material changes its properties as a result of operational loads action. There are situations when construction leaves its working state with satisfactory material characteristics. This is due to the process of formation and development of cracks as structural elements. The article proposes to analyze history of the development process, which makes it possible to study mechanisms of its flow and to determine factors of process control. The analysis of the studies carried out on the effect of prescriptional and technological factors on the properties of materials and products, the phenomena of self-organization and the role of cracks in materials and products showed that the emergence of critical events cannot be foreseen when assessing properties accepted by averaged characteristics. The construction is considered to be a complex self-organizing system. This determined the work aim: to justify the choice of dynamic models, allowing to determine the structure elements, interaction of which ensures functional state of the structure. The qualitative method of verbal description of the structure of systems with the use of cognitive approaches was laid in basis of adopted methodology. Conservative, metastable and active elements are distinguished in the construction-system. It is emphasized that the active history of the system is determined by active elements, to which cracks refer to. Cracks are considered as some necessary structural elements, present at all scale levels. It is shown that the formation of cracks-systems, treated as parasitic systems, terminates the end of construction-system development and the beginning of parasite system development. Completion of crack-system life cycle removes the construction-system from its working state. To get constructions with a given set of structural elements, a creative community of materials and designers is needed, which will create a common terminology dictionary for both specializations.

1. Introduction

Construction products and structures in the period of operation perceive a complex of normalized and non-normalized power loads and loads connected with environmental effects (seasonal and diurnal changes in temperature and humidity). Under the influence of such influences, the properties of materials change, which inevitably affects the properties of products and structures. The level of change in the properties of materials is regulated by regulatory documents, which allows you to quantify the current state of the structure. The accumulated experience has shown that with satisfactory average values of the required material properties, a critical situation can arise in the design. As a rule, critical events are localized and are associated with the formation of a new component for the material and construction of the component – a crack. The process of crack development is initiated by external and internal factors and provokes structural rearrangements in the material and the redistribution of stress and strain in the structure, which can stimulate further deepening of the process of crack development to critical parameters. Any process, including the process of crack growth, cannot be analyzed by stopping it. Therefore, in order to establish the management factors, it is necessary to study the history of the development of the process, which will allow us to develop an idea of the mechanisms of its flow taking into account the time factor.

The compulsory presence of the history of development makes it possible to trace the structural transformation of the material and product taking into account that there are no "good" or "bad" elements – they are all parts of the whole. Representation of the material and construction in the form of an inseparable integrity allows us to consider the design as an open complex self-organizing system [1,2,3]. The system does not exist outside its structural design. Professor Solomatov V.I. and his followers emphasized the structural structure of composite building materials and products. The interrelation and mutual influence of different levels of structural components is noted in the works of Konokhov P.G., Krivenko P.V., Bolshakov V.I., Derevyanko V.N., Plugina A.N. and Plugin A.P., Runova R.F., Vandalovsky A.G., Bratchun V.I., I. Shtark, Fainer M.S., Shinkevich E.S. and many other native and foreign researchers.

The processes of self-organization during the formation of the structure of exploitation of building composites are devoted to the works of Korchakava M.D., Seynich L.A., Pushkareva E.K., Barabash I.V., Bobryshov A.N., Khrulev V.M., Chernyavsky V.L., S. Fits and others. The role of cracks in the formation of the properties of materials and structures is shown in the studies of Karpenko N.I., Piradov K.I., Zaitsev I.V., Dorofeev V.S., Bolotin V.V., Grushko I.M., Luchko I.I., Solodky S. I., Leonovich S.N., R. Lermit. The influence of prescription and technological factors on the formation of the properties of building materials and products has been studied in the studies of Voznesensky V.A., Lyashenko T.V., Dvorkin L.I. and Dvorkin O.L., Koval S.V., Sheinin A.E., Akhverdov I.N., Sanytsky M.A., Gorchakov T.I. and many other researchers.

All of the above studies (and many others) are carried out, as a rule, within the framework of the current paradigm, which emphasizes that the level of the final properties of materials and products is determined by the initial composition and technological conditions for obtaining materials and their processing into a structure. In the final design, the material is considered as a kind of given outside the history of its formation and outside the specifics of its structural design. The mismatching of the terminology of technologists-materialists and designers does not contribute to the development of joint configurations of interactions. It is obvious that there are no products and structures outside their material filling, as there is no material without its constructive design. The system approach allows organically to combine material and construction into an inseparable whole – a system with a given objective function. In such a system, the principle of complicity of all structural elements in the functionally oriented behavior of the structure-system is realized. When analyzing a structurally structured system, it is possible to create a common terminological dictionary that will mutually expand the conceptual apparatus of materials scientists and designers and create the prerequisites for other basic models that shape our professional and scientific views and beliefs.

To implement the declared approach, a task was identified on the justified choice of dynamic models, allowing to determine a set of elements of the structure, the presence, interaction and development of which ensures the operational functional state of the system design.

2. Research methodology

In the initial description and analysis of the construction-system, the verbal method of description [4] is appropriate. It allows to reveal the structure of the system at a qualitative level, to establish the basic processes of its development, to describe the mechanisms of interaction between different-scale structures. In turn, the verbal method is basic in the implementation of cognitive approaches, which aim to investigate the processes of perception, categorization, comprehension and explanation of objects, as well as the representation and storage of knowledge [5,6]. The cognitive approach is appropriate in cases: – if the interconnectedness of the processes leads to the need to consider individual phenomena in their totality; – in the absence of sufficient quantitative information on the dynamics of structural development; – with the development of self-organization processes, which makes it difficult to build quantitative models. The use of methods of cognitive modeling, based on the conduct of thought experiments, allows, with the accumulation of relevant information, to move to quantitative models. In this paper, methods of qualitative description of the construction-system are used. Previous studies [3,7] allowed to propose a classification of structural components, based on the different "sensitivity" of different elements under the influence on the design-system of external and internal factors. In the general case, it is possible to distinguish conservative, metastable and active structural elements.

Conservative (from Latin conservation – preservation) structural elements include elements that change their parameters slowly enough during the period when the object performs its functions. As a rule, conservative elements belong to a scale level on which gravity dominates and on which phenomenological approaches to property evaluation are realized.

Elements of the structure that have relative stability under given conditions and that are capable of spontaneously transitioning to a stable or unstable state under the action of external or internal factors are referred to metastable (from the Greek meta – between, after, through) the elements of the structure.

A distinctive feature of active (from Latin activus – active) elements of the structure is their ability to respond adequately to the effects tested in a commensurate time interval (one tempo-rhythm).

The entire set of structural elements system in the form of building products and structures acquires in the technological period of obtaining concrete and its processing into products. At this stage all the initial elements actively participate in the processes of a different-scale structural design with the formation of new elements of the structure of the emerging system. Situations in which the emerging intermediate structural formations initiate the initiation and development of fundamentally different elements that are part of the structure of the final product are not ruled out. Characteristic manifestations of the effects of equifinality by which in our case we mean the implementation of various mechanisms of structural organization, which lead to the appearance of qualitatively identical elements of the structure. As examples, the origin of technological cracks is considered: – when discrete structures are formed at the level of products of neoplasms and at the level of interaction of the initial grains of mineral binders; – with the development of gradients of intrinsic deformations at interfaces between hardening matrix material and aggregates; – under the action of emerging gradients of deformation at the level of products or structures.

Thus, during the functioning of the system, the system enters with a certain set of structural elements, including the presence of technological (residual, initial, hereditary) local and integral deformations. Therefore, functioning should be considered as an uninterrupted structural drift in the direction in which the effects of adaptation of the system are manifested in the conditions of the permanent impact of the operational environment. In this regard, the life cycle of the structure is proposed to be described by the history of the evolution of the structural components, Fig.1.

The co-evolution of elements of the structure that are different in form and purpose presupposes their interaction, which is the primary basis of stability-the ability of the system to maintain a dynamic equilibrium through the manifestation of internal and external security.

Particular attention should be given to cracks. The term "crack" common to materials scientists and designers, initially carries a negative connotation. A crack is a localized structure element. The phenomenon of localization leads to the disintegration of a continuous medium into separate local structures. Therefore, when a "crack" appears in the construction, then automatically the paradigm of continuity loses its meaning. Our researches give grounds to assert that cracks are an inevitable and necessary structural component at all levels of structural heterogeneities. Cracks-builders contribute to strain relaxation, including critical retain properties of the whole by changing the parameters of its own, can be transformed into the inner surface section, include a work metastable and conserved structural elements.

Interaction and mutual influence of structural elements increases the diversity of structure, contributes to the development of interstructural links, thus providing a manifestation of the phenomena of homeostasis, which ensures the viability of the system. To realize the self-movement of the structure towards the self-preservation of the system-design, it is necessary to determine the priority series of elements capable of influencing the periods of the onset of unfavorable situations. This priority number ranks first crack, which for some reason has moved from rank-creator-destroyer crack, thereby implementing the idea of self-organized criticality [8] and determine the start of a history of the life of the system period.

From the moment of formation of the crack-destroyer (point M, Fig. 2), it is necessary to consider two histories of development of two systems – the basic design-system and the newly formed crack-system. The crack-system, by virtue of its activity, which manifests itself through the redistribution of deformations with their concentration along the crack front, the release of surface energy as the area of the fracture edges increases, various-scale cracks and interfaces become involved in its development, reoriented the functions of the basic system to itself.

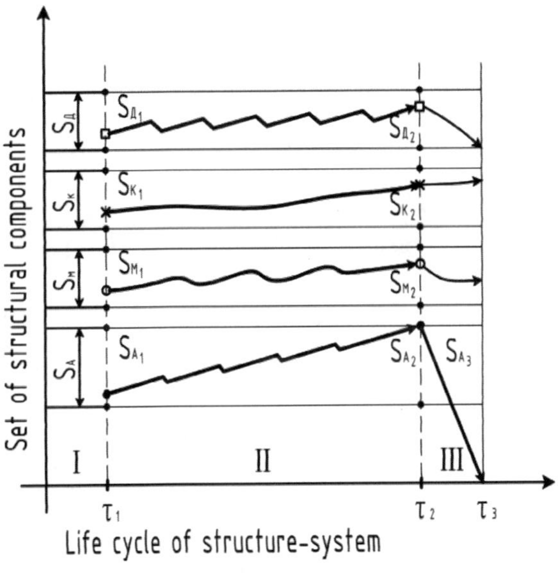

Fig. 1. Co-evolution of the elements of the structure in the period of functioning of the construction-system

I – stage of formation of the system-design (τ_1); II – period of functioning of the system-design ($\tau_1 \ldots \tau_2$); III – the stage of the output of the structure-system from the safe functioning ($\tau_2 \ldots \tau_3$)

S_A, S_M, S_M, $S_Д$ – structural components of the construction-system;

S_{A1}, S_{M1}, S_{K1}, $S_{Д1}$ – initial set of active (A_1), metastable (M_1), conservative (K_1) elements and technological deformations ($S_{Д1}$);

$S_{A1} \ldots S_{A2}$ – change the parameters of active elements;

$S_{M1} \ldots S_{M2}$ – change the parameters of metastable elements;

$S_{K1} \ldots S_{K2}$ – change of parameters of conservative elements;

$S_{Д1} \ldots S_{Д2}$ – change of local and integral residual deformations;

S_{A3} – change in the structural parameters of the self-developing parasitic system.

With an increase in the degree of structural diversity of the design, the level of its safe functioning decreases. During this period, the behavior of the design-system is determined by the dynamics of the development of the new system, which is referred to the parasitic system.

This definition is legitimate in our view, since the fracture self-generated in the basic structure and draws energy for its own development from its environment – the material of the system design.

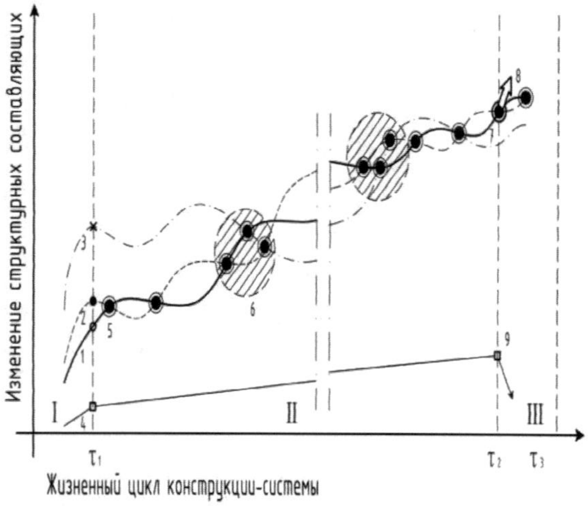

Fig. 2. Dynamic model of changing structural diversity over the lifecycle of the design-system

O – the minimum acceptable level of safe operation;
A – the necessary set of structural elements, ensuring the safe functioning of the structure-system;
(A-B) – structural transformations in the period of functioning;
B – spontaneous generation in the construction-system of the parasite system;
(B-C) – integral structural changes of the base system;
M – reorientation of the objective function of the structure-system to the goal of the self-generated system;
(C-D) – reducing the level of safe operation of the system-design;
(M-N) – active functioning of the parasite system;
N – completion of the life cycle of the parasitic system;
D – completion of the life cycle of the construction-system.

The purpose of the existence and development of a parasitic system in the complete passage of its own stages of its life cycle – from spontaneous generation through development to the completion of its existence. The completion of the existence of a crack-system is the exit of its front line to the surface of the base system. This concludes the life cycle of the parasite system, which causes the inevitable completion of the life history of the structure-system. Thus, the inflation of the structural-system structure through degeneration of the whole variety of structural components before the emergence of a new relatively simple parasite system leads to a violation of its external security and a complete exit from the functional state.

3. Summary

3.1 The presentation of building products and structures in the form of structurally organized systems at this stage of research should be regarded as a peculiar beginning of the explanation of the mechanisms of the functioning of structured objects.

3.2 This structural approach is based on an entrenched scientific ideology based on ideas and methods of systemic thinking and synergetics, for which the concept of "structure" is dominant.

3.3 The inevitable shift of paradigms towards structured objects will be greatly accelerated when creating a joint vocabulary of technologists and designers and developing a common strategic direction for obtaining system designs with the required set of necessary structure elements.

The analysis performed allows to conclude that building constructions should be considered as an independent research object due to their individual structural design, which is defining for the manifestation of the required properties. Special attention should be paid to active elements since a change of their parameters causes change of parameters of the construction structure. All stages of the construction life cycle are accompanied by the manifestation of deformations, included as an element in the general network of the processes of its creation and functioning. The propagation of deformations in the material is carried out in the form of returnable deformation waves, which are transmitted from one level of heterogeneities to another through the surfaces of partition between them. That promotes to the spontaneous generation and self-development of active elements – the technological cracks and inner surfaces of partition that determine material damage. Experimental researches allowed recommending methods for controlling the formation and propagation of deformations in the structure of the material of construction at various levels of heterogeneities. That provided the possibility of the predicted damage change as set of the active elements which define maintaining integrity and manifestation of material properties in the construction.

References

[1] Vyrovoy V, Dorofeev V, Sukhanov V 2010 *Composite Building Materials and Constructions. Structure, Self-organization, Properties* (TESS, Odessa)

[2] Prangishvili I 2000 *System Approach and System-wide Regularities* (SINTEG, Moscow)

[3] Sukhanov V, Vyrovoy V, Korobko O 2016 *Structure of Material in Structure of Construction* (POLYGRAPH, Odessa)

[4] Solomatov V, Vyrovoy V, Bobryshev F and others 1991 *The Polystructural Theory of Composite Building Materials* (FAN, Tashkent)

[5] Cherniavsky V 2008 *Adaptation of Abiotic Systems: Concrete and Reinforced Concrete* (DNURT, Dnepropetrovsk)

[6] Dorofeev V, Vyrovoy V 1998 *Technological Damage of Building Materials and Constructions* (Gorod Masterov, Odessa)

AUTHOR INDEX